London '72 

PELICAN BOOKS

WE ARE NOT ALONE

Walter Sullivan, science editor of the *New York Times*, was born in New York on 12 January 1918. He joined the staff of the *Times* as a copy boy in 1940 after graduating from Yale University, and returned there in 1946, first as a reporter and then as a foreign correspondent. He has won a number of awards, among them the non-fiction prize for *We Are Not Alone* in 1965. He has contributed to, and appeared on, American TV and also appeared on French and English radio and TV programmes. His publications include: *Quest for a Continent* (1957), *Assault on the Unknown* (1961), *America's Race for the Moon* (1962; editor and contributor), and *The Scientific Study of Unidentified Flying Objects* (1969; Introduction). He was also co-author of *Project Apollo*, a report on America's effort to land a man on the moon originally published in the *Times*, and of *The Soviet Union. The Fifty Years*, a similar effort. Mr Sullivan belongs to a number of societies and associations. He has recently been made Doctor of Humane Letters at Yale University.

WALTER SULLIVAN

We Are Not Alone

PENGUIN BOOKS

Penguin Books Ltd, Harmondsworth, Middlesex, England
Penguin Books Australia Ltd, Ringwood, Victoria, Australia

—

First published in the U.S.A. 1964
Published in Great Britain by Hodder & Stoughton 1965
Revised edition published in Pelican Books 1970

—

—

Made and printed in Great Britain
by Hazell Watson & Viney Ltd
Aylesbury, Bucks
Set in Monotype Times

To those everywhere
who seek to make
'L'
a large number*

*See page 280

Contents

Preface

BECAUSE this book is written for those with a minimum background in science, the arguments are presented in a manner that some specialists will find elementary. For those who wish to pursue the subjects further, references for each chapter are presented at the back of the book. In the interest of brevity, titles such as 'Doctor' or 'Professor' are omitted. Temperatures, except where otherwise noted, are in degrees Fahrenheit. The expression 'billion' is used in the American sense of the word, meaning 1,000 million.

Acknowledgements

ALL chapters of this book dealing with scientific matters have been reviewed by specialists in an effort to avoid inaccuracies and misrepresentations. The author has sought to incorporate their suggestions in so far as possible. However, several chapters deal with contemporary controversies that are complex and, since this book is intended chiefly for laymen, no attempt has been made to set forth all arguments on each side. Rather, to communicate the flavour of the debates, only those aspects most readily understood have been included.

The following have generously reviewed portions of the manuscript within their special areas of competence: Edward Anders, Associate Professor of Chemistry, Enrico Fermi Institute for Nuclear Studies, University of Chicago; Melvin Calvin, Professor of Chemistry, University of California at Berkeley; A. G. W. Cameron, Institute for Space Studies, Goddard Space Flight Center, National Aeronautics and Space Administration; Frank D. Drake, Jet Propulsion Laboratory, California Institute of Technology; Christian T. Elvey, University Research Professor, University of Alaska; Su-Shu Huang, Goddard Space Flight Center, NASA; Peter van de Kamp, director, Sproul Observatory, Swarthmore College; Brian H. Mason, Curator of Mineralogy, American Museum of Natural History; Stanley L. Miller, Associate Professor of Chemistry, University of California, San Diego (La Jolla); Philip Morrison, Visiting Professor of Physics, Massachusetts Institute of Technology; Bartholomew Nagy, Professor of Chemistry, Fordham University; Edward M. Purcell, Professor of Physics, Harvard University; Freeman H. Quimby, Chief, Exobiology Programs, Bioscience Programs Division, NASA; Carl Sagan, Smithsonian Astrophysical Observatory; Harlow Shapley, Emeritus Director, Harvard College Observatory; Lyman Spitzer, Jr, Director, Princeton University Observatory; Kai Aa. Strand, Scientific Director, U.S. Naval Observatory; Charles H. Townes, Provost, Massachusetts

Institute of Technology; Harold C. Urey, Professor-at-Large of Chemistry, University of California, San Diego (La Jolla); and Gérard H. de Vaucouleurs, Associate Professor of Astronomy, University of Texas.

It is not possible to cite all who have assisted in furnishing material and in other ways. Among those who were especially helpful in this regard were: Leland I. Anderson of Minneapolis; Giuseppe Cocconi of CERN, the European Organization for Nuclear Research near Geneva; Sidney W. Fox, Director, Institute for Space Biosciences, Florida State University; Lancelot Hogben, Vice-Chancellor and Principal, University of Guyana, British Guiana; Norman H. Horowitz, Professor of Biology, California Institute of Technology; John A. Kessler, Lincoln Laboratory, Massachusetts Institute of Technology; Gerard P. Kuiper, Director, Lunar and Planetary Laboratory, University of Arizona; Fr. Clement J. McNaspy, S.J., Associate Editor, *America*; Alfred P. N. F. Stiernotte, Professor of Theology, Theological School, St Lawrence University; Alexis N. Tsvetikov, Department of Biophysics, Stanford University; and Richard S. Young, Chief, Exobiology Division, Ames Research Center, NASA.

My foremost debt is to Frank D. Drake, for advice and assistance, and to Philip Morrison, whose lectures on this subject were the initial inspiration for this book.

Particularly valuable editorial assistance was rendered by Elizabeth Burpee, Patricia Wohlgemuth and my wife, Mary. Of inestimable help were the research facilities of the New York Public Library, particularly its Science and Technology Division, and those of thc *New York Times*. A considerable portion of the material in this book was collected in the service of the *Times* and my thanks are due to that newspaper for permission to draw upon this material and on art work used to illustrate it. I am also grateful to the following for permission to use previously published material:

Drawings: *Fortune*, for several drawings from the March 1961 issue, reprinted by courtesy of *Fortune Magazine*; Dr John Hall, Director, Lowell Observatory, Flagstaff, Arizona, for drawing by Percival Lowell; Harvard University Press for asteroid orbits taken from *Between the Planets* by F. G. Watson, rev. ed., © 1956; Dr Su-Shu Huang for drawing used in his article in the June 1961 issue of *Sky and Telescope*, and for two drawings used with his April 1960 article in *Scientific American*; *International Science and Technology* for two drawings from their October 1962 issue, pp. 55–60; Lincoln Laboratory, M.I.T., for diagram of astronomical distances; Milton K. Munitz for two drawings from *Theories of the Universe*, Free Press, © 1957; *Scientific American* for two drawings from the April 1960 issue, reprinted with permission; *Sky and Telescope* for the drawing in the June 1961 issue and for the diagram of Angular Momentum in the solar system, published in the January 1960 issue; George Philip & Son, Ltd, The London Geographical Institute, for star finder used as basis of a drawing; *Physics Today* for drawings on the origin of the solar system with an article by T. L. Page, vol. 1 (1948), no. 6, pp. 14–15. A number of the drawings were prepared by Andrew Sabbatini of the *New York Times*.

Photographs: Bell Telephone Laboratory; Jet Propulsion Laboratory; Lick Observatory; Mount Wilson and Palomar Observatories, U.S. Naval Observatory.

Text: *Of Stars and Men* by Harlow Shapley, Beacon Press, © 1958; *Interstellar Communication*, by A. G. W. Cameron, W. A. Benjamin, Inc., © 1963; *The Origin of Life* by Alexandr Ivanovich Oparin, translated by S. Morgulis, published by Dover Publications, Inc., and reprinted through permission of the publishers, published in Britain by Oliver & Boyd, Edinburgh. *Mars* by Percival Lowell, Houghton Mifflin Co., 1895; *Louis Pasteur, Free Lance of Science* by René J. Dubos, © 1950, by René J. Dubos, reprinted by permission of Little Brown & Co.; *God and the Astronomers* by W. R. Inge, Longmans, Green, London, 1933; *Lincos* by Hans Freudenthal, North-Holland Publishing Co., Amsterdam, 1960; *Saddharma-Pundarika* in *Sacred Books of the East*, vol. 21, translated by H. Kern, Oxford University Press; *On the Nature of the Universe* by Lucretius, translated by Ronald Latham, Penguin Books, © 1951; *The Origins of Life, New Biology* No. 16 by J. B. S. Haldane, Penguin Books, © 1954; *Poems*, complete edition, by Alice Meynell, Chas Scribner's Sons, 1923.

1

Order of the Dolphin

In November 1961 the most august scientific body in the United States convened a meeting at the National Radio Astronomy Observatory in Green Bank, West Virginia. While it was not held in secret, in the official sense, every effort was made to avoid publicity because of the sensational nature of the question to be discussed.

The subject was 'Intelligent Extraterrestrial Life'. Yet this was no gathering of wild-eyed dreamers. The convener was the Space Science Board of the National Academy of Sciences. The host was one of the world's most distinguished astronomers, and several others among the eleven men present were internationally recognized leaders in their highly diverse fields. In fact, J. P. T. Pearman of the Space Science Board smuggled three bottles of champagne from Washington, D.C., anticipating that one participant might win a Nobel Prize during the conference. This did, indeed, happen, throwing the sessions into somewhat of an uproar.

Despite their diversity, the subjects in which the conferees were expert all bore, in some way, on the problem of whether there is intelligent life elsewhere than on earth and, if so, how to communicate with such beings. The participants shared a strong feeling that such civilizations exist. They included optical astronomers, radio astronomers, a physicist, an astrophysicist, a biochemist, an investigator into chemical steps whereby life may have originated on earth and specialists in communications, both among animals and among men. Because one of them, J. C. Lilly, specialized in the 'language' of dolphins, the group lightheartedly constituted itself 'The Order of the Dolphin', and subsequently its Nobel laureate, Melvin Calvin, had buttons made showing a dolphin emblem from an ancient Greek coin. He sent one to each participant.

These men were aware that some of their colleagues would

scoff when they learned of this discussion, and they sought, insofar as possible, to bring the disciplines of the scientific method to bear upon the question. They worked out a mathematical equation designed to indicate the number of civilizations that may have evolved to the point where they can – and want to – communicate with other solar systems. The conclusion of the conferees was that the number may be large, although no reliable estimate is yet possible. Such civilizations, some of them felt, may long have been watching our solar system for the first signs of an awakening technology; their calling signals may, at this very moment, be impinging on the earth, the question being how and where to look for them.

Only a few years ago to raise such questions would, so far as scientific reputation is concerned, have been suicidal. Yet in 1959 it was proposed that the giant, federally owned, radio telescope at Green Bank be aimed at two nearby stars to see if, by any chance, intelligent signals were emanating from them. The effort, Project Ozma, was carried out shortly thereafter, and was a prelude to the conference that followed.

As anticipated, the use of an instrument as important as the Green Bank radio telescope for such a project was viewed askance by some scientists. Sir Bernard Lovell, director of what was then the largest instrument of this kind, at Jodrell Bank in England, was at first among the sceptics, but he modified his views. On 22 March 1962 he was asked by Emilio Q. Daddario, congressman from Connecticut in the House Committee on Science and Astronautics, what he thought the possibilities were for receiving signals from other planets and what kind of programme, if any, the United States should undertake to intercept them.

'Well sir,' Sir Bernard replied, 'I think that now one has to be sympathetic about an idea which only a few years ago would have seemed rather far-fetched.' Harrison S. Brown, Professor of Geology at the California Institute of Technology and high in the scientific councils of the federal government, echoed Sir Bernard. He said to the committee: 'I believe, as to the question of extraterrestrial life, that it is one of the most important and exciting problems that confront us.'

A somewhat similar evolution of thought occurred in the other country rich and powerful enough to do something about this question. At the time of Project Ozma an American asked one of the leaders in Soviet radio astronomy if the Russians were planning any such project. He confessed his own interest in the subject, but said, 'If I proposed anything of that sort, the government would think I was out of my mind.'

At the end of 1962, however, the Soviet Academy of Sciences published an entire book on the universality of intelligent life, written by Iosif S. Shklovsky, one of the most brilliant theoretical radio astronomers alive. While the book was issued as part of the academy's 'popular science series', Shklovsky's treatment of the subject was aimed at his peers as much as at the scientifically oriented layman. This was followed, in the autumn of 1964, by a conference on 'extraterrestrial civilizations', held at the Burakan Astrophysical Observatory, one of the foremost such institutions in the Soviet Union. The most dramatic Soviet development was a report by one of Shklovsky's associates early in 1965 that CTA-102, an object emitting radio signals from space, might be a supercivilization seeking to draw attention to itself.

Why, in the second half of the twentieth century, has this subject so abruptly become 'respectable', after being so long corrupted and discredited by various science-fiction writers, comic strips and, more recently, television shows? To understand how this came about it is necessary to retrace the evolution of man's concept of the cosmos, and his place in it, that has occurred since our ancestors first began to wonder about such things.

2

Spheres within Spheres

BECAUSE of the limits that nature places upon our vision, our view of the world until we are told otherwise is of a flat land beneath the changing vault of the heavens. So is it in our childhood and so was it for primitive man. Yet, long before even our most ancient myths were born, man had begun to speculate about this environment. Such thinking was one of the essential qualities that made him man. He early saw the heavens as a rotating sphere of fixed stars across which moved the seven 'wanderers': sun, moon and the five planets visible to the naked eye.

For most of human history it has been thought that the stars were fixed to some sort of sphere or were pinholes in that sphere, permitting glimpses of a universal fire that flamed beyond. Whatever the 'model' of the universe, man was at the centre – quite natural in view of the fact that the celestial sphere seems equidistant in all directions.

The concept of our central position has been hard a-dying. It is a mark of the climactic period in which we live that, after thousands of years of sophisticated observation and speculation, it has only now perished. Some of those who finally laid it to rest (for example, by showing the true, off-centre location of the sun within our galaxy) are still alive. After centuries of resistance, theologians in some of the major Christian denominations have begun to grapple with the religious implications of these discoveries – in particular the fact that not only are we not central in the scheme of things, but we may be inferior, physically, mentally and spiritually, to more highly evolved beings elsewhere.

As Harold P. Robertson, Professor of Mathematical Physics at the California Institute of Technology, has put it, early man 'sat serenely in the middle, the favoured onlooker if not the master of all he surveyed.' Today he finds himself resident on a planet orbiting an average star in an average galaxy, confronted by the strong possibility that his intelligence – and virtue – are inferior to those in at least some other solar systems.

Almost as soon as men first began to speculate, scientifically, on the nature of the objects that they saw in the sky, some recognized that there may be other worlds like our own. The first flowering of astronomical interest began towards the end of the seventh century before Christ, and the power of objective reasoning in this area was dramatically illustrated on 28 May 585 B.C. (Julian Calendar), during a battle between the Medes and Lydians. As the two armies fought, the sun was suddenly obscured, even though the sky was clear. The stunned armies ceased fighting and peace was concluded. When the ancient world learned that this eclipse had been predicted shortly beforehand by Thales of Miletus (a Mediterranean port in what is now Turkey), the study of celestial bodies took on new meaning.

Thales is said to have proposed that the stars are other worlds, but it was his pupil, Anaximander, who appears to have been the first to elaborate the idea that the number of worlds is infinite, some of them in the process of birth, some dying. The earth, cylindrical in shape, stands without need for support at the centre of this universe, he said, and the sun is as large as the earth (an early hint of the true dimensions of the solar system). Like many of the prophetic concepts that came in subsequent centuries, the idea that there may be many worlds was too far ahead of its time to be widely accepted. Even the view of Pythagoras, a young contemporary of Thales and Anaximander, that the earth is spherical, while it was accepted by many ancient Greeks and by the Arabs, was largely extinguished until the Renaissance.

One of the difficulties was that such concepts often were at variance with the religious myths and dogmas of that time. Another contemporary of Pythagoras, the philosopher Xenophanes of Colophon, ridiculed the anthropomorphic gods worshipped by his Greek countrymen, noting that cows, if they could make graven images, would worship bovine gods. In seeking to put man in his place he, too, postulated worlds unlimited in number, though not overlapping in time, and said the moon is inhabited. The same views were expressed by Anaximenes of Lampsacus, friend and companion of Alexander the Great, who told that warrior, to his astonishment, that he had conquered only one of many worlds.

One of Alexander's friends had trouble with another protagonist of multitudinous worlds. The latter was the philosopher Zeno of Elea, who plotted the overthrow of Nearchus, Macedonian general and ally of Alexander. After his arrest Zeno mischievously named some of the tyrant's closest followers as co-conspirators and then insisted that he had even more important information to whisper into the ear of Nearchus himself. When permitted to do so, he bit the ear and held on until stabbed to death, which, perhaps, is evidence that the advocates of extraterrestrial life have been rebels at heart from the beginning.

In this respect, one of the most radical innovators in the philosophical and scientific thought of ancient Greece was Democritus, inventor of the theory that all matter is made of 'atoms' – indivisible particles too small to be seen, indestructible and eternal. The earth was formed, he said, from a whirling mass of these atoms and, since space and time are infinite and the atoms are for ever in motion, there must now be, and always have been, an infinite number of other worlds in various stages of growth and decay. His description of natural phenomena, written in the fourth century B.C., avoided any recourse to the supernatural.

His views, as modified by Epicurus, were magnificently set forth 350 years later by the Roman poet Lucretius, in his *De Rerum Natura* (*On the Nature of Things*). Lucretius opened with a ringing tribute to Democritus:

> When human life lay grovelling in all men's sight, crushed to the earth under the dead weight of superstition whose grim features loured menacingly upon mortals from the four quarters of the sky, a man of Greece was first to raise mortal eyes in defiance, first to stand erect and brave the challenge. Fables of the gods did not crush him, nor the lightning flash and the growling menace of the sky. . . . He ventured far out beyond the flaming ramparts of the world and voyaged in mind throughout infinity. Returning victorious, he proclaimed to us what can be and what cannot . . . superstition in its turn lies crushed beneath his feet, and we by his triumph are lifted level with the skies.

Lucretius' presentation of the ancient atomic theory sounds strikingly familiar to any student of modern physics and quantum mechanics. It also expresses a view that still strongly influences those today who believe life exists elsewhere than on earth, namely that no phenomenon in nature, including the emergence of life on a planet, occurs only once.

> Granted, then [he wrote], that empty space extends without limit in every direction and that seeds innumerable in number are rushing on countless courses through an unfathomable universe under the impulse of perpetual motion, *it is in the highest degree unlikely that this earth and sky is the only one to have been created* and that all those particles of matter outside are accomplishing nothing. This follows from the fact that our world has been made by nature through the spontaneous and casual collision and the multifarious, accidental, random and purposeless congregation and coalescence of atoms whose suddenly formed combinations could serve on each occasion as the starting-point of substantial fabrics – earth and sea and sky and the races of living creatures. . . . You have the same natural force to congregate them in any place precisely as they have been congregated here. You are bound therefore to acknowledge that in other regions there are other earths and various tribes of men and breeds of beasts.

Nothing in the universe, he said, 'is the only one of its kind, unique and solitary in its birth and growth.' As there are countless individuals in every species of animal, from 'the brutes that prowl the mountains, to the children of men, the voiceless scaly fish and all the forms of flying things', so there must be countless worlds and inhabitants thereof.

During the lifetime of Lucretius and the century that followed, this concept spread far and wide. Plutarch in his *De Facie In Orbe Lunae* (*Regarding the Face of the Moon's Disc*) discussed the habitability of the earth's great satellite – whether its lack of clouds and rain mean it is intolerably dry, and so forth. Elsewhere he wrote of other theories, including one in which the universe is triangular with sixty worlds on each side and one at each apex, making a total of 183 inhabited earths. A picture similar to that of Lucretius was presented in China by Teng Mu, a scholar of the Sung Dynasty, who wrote:

Empty space is like a kingdom, and earth and sky are no more than a single individual person in that kingdom.

Upon one tree are many fruits, and in one kingdom there are many people.

How unreasonable it would be to suppose that, besides the earth and the sky which we can see, there are no other skies and no other earths.

This belief rested primarily on abstract reasoning and, while it was attractive to philosophers, its general acceptance, until after the Renaissance, was prevented by the enclosure of our world in a solid sphere, forged by the earliest astronomers. This was the firmament to which the stars seemed to be fixed, rotating around the earth approximately once a day and traversed by the sun, moon and five visible planets. The chief problem of ancient astronomy was to explain the motions of these wanderers and thus make possible accurate predictions of their movements.

Pythagoras and his followers believed all of these bodies, including the earth, move around a central fire that is for ever on the far side of the earth and hence invisible to man. Sunlight, they said, was simply a reflection of this fire. Anaxagoras then identified the sun as a huge, red-hot stone and the moon as a cool body, illuminated by sunlight like the earth, and also inhabited. The Milky Way, he said, was a portion of the celestial sphere shielded from sunlight by the earth. However, despite these advances he clung to the idea that the earth is flat.

He also ran into the bigotry that so often has beset astronomers. This was the age of Pericles, a time of glorious creativity in Athens, but also a time of embitterment by military defeat and suspicion of 'foreign' ideas. The teachings of Anaxagoras, so out of line with classic Greek religious concepts, were deemed impious and subversive. He was tried and, it is said, was to be executed when his friend Pericles intervened. Socrates, reportedly the pupil of Anaxagoras, was not so fortunate.

The efforts of the early astronomers to account for the movements of the wanderers failed because none realized that all the planets, including the one from which we gaze upon the heavens, are travelling in elliptical paths around the sun. Without this knowledge the motions of the sun, moon and planets appear most peculiar. In fact at times some planets seem to reverse direction and move backward against the panoply of stars.

In the early days of their astronomical studies the Greeks had only such elementary tools as trigonometry at their disposal, although by the third century B.C. they had advanced to where they could calculate the areas of complex curved surfaces. The simplest and most 'perfect' curved surface was, of course, that of a sphere, and this set in motion a train of thought that was not broken for centuries. It plagued all attempts to explain the movements of celestial bodies until the time of Kepler.

Plato, the most noted pupil of Socrates, wrote that God 'made the world in the form of a globe, round as from a lathe, having its extremes in every direction equidistant from the centre, the most perfect and the most like itself of all figures . . . and He made the universe a circle moving in a circle, one and solitary, yet by reason of its excellence able to converse with itself, and needing no other friendship or acquaintance.'

Plato's idea of a perfect universe, with the world at its centre, led him to believe there must be only one world. His concept that such perfection was manifested in circles and spheres led his contemporary, Eudoxus of Cnidus, to explain the movements of the celestial bodies in terms of their 'attachment' to a series of theoretical spheres. Although complex, his theory was simple compared to those that followed. The stars, as already accepted, were fixed to an outer sphere that rotated uniformly around the earth once in about twenty-four hours. However, to account for the seemingly erratic movements of a planet such as Jupiter, he proposed that it is fixed to the equator of a rotating sphere whose poles are, in turn, fixed to the equator of another sphere, rotating in a different direction. That sphere, in turn, is twisted by the rotary motion of a third sphere, and so forth. The combination of these motions was sufficient, in a very approximate way, to account for the observed movements. All told, the theory of Eudoxus called for twenty-seven spheres: one for the stars, three each for the sun and the moon, and four each for the five known planets.

While Eudoxus considered these spheres mathematical devices, his contemporary Aristotle said they were solid, though transparent, objects. In an effort to account for the motions more precisely he proposed fifty-five of them. Another man of that time, Heraclides of Pontus, a pupil of Plato, explained the movement of the stars by saying it is the earth that turns, spinning eastward to make one revolution per day. To account for the changes in brightness of Mercury and Venus, he proposed that they circle the sun, which in turn circles the earth. Because those two planets lie nearer the sun than does the earth and orbit faster, their variations in brilliance are particularly noticeable.

Five years after the death of Heraclides a man was born on the island of Samos, birthplace of Pythagoras, who came to be known as 'the ancient Copernicus', for he proposed that the sun – not the earth – is at the centre of the universe. The planets and a spinning earth move in circles around the sun, he said. This man of vision, 1,800 years ahead of his time, was Aristarchus of Samos. His idea that the sun must be at the centre may have arisen from his calculation that it is some 300 times bigger than the earth and eighteen to twenty times more distant than the moon.

We do not know if he made any attempt to measure actual sizes and distances, but his contemporaries were making bold – and remarkably successful – attempts to calculate the size of the earth. This figure could then be applied to the geometry of eclipses to estimate actual, rather than relative, distances and sizes for sun and moon. The best-known of these calculations was that made by Eratosthenes, head of the great library at Alexandria, based on observations at the summer solstice in 284 B.C. On that day, every year, he knew that the full image of the sun could be seen at the bottom of a deep well near the edge of the Libyan Desert at Syene, present site of the Aswan Dam on the Nile. Thus the sun at Syene was directly overhead, whereas the shadow cast by a stick at Alexandria showed the angular distance of the sun from the zenith to be one-fiftieth of a full circle. Since Eratosthenes believed Alexandria to be due north of Syene and about 5,000 stades distant, over the curved surface of the earth, he calculated the total circumference of the earth to be 50 × 5,000, or 250,000 stades. He later refined this to 252,000 stades. If one uses Pliny's version of the type of stade to which Eratosthenes referred, this works out to a circumference of 24,662 miles, compared to an actual distance, through the poles, of 24,860 miles. While the accuracy was largely luck, the calculation gave the ancient world an idea of the size of our planet. When this was applied to the calculations of distance to sun and moon, in terms of the earth's diameter, the vast size of the solar system became apparent.

As the glory of Greece faded, there were these glimpses of the true scale of our corner of the universe, as well as far-reaching philosophical speculations regarding the existence of other universes or other worlds. But the celestial spheres were confining, and bold inquiry into such matters was on the decline. Furthermore, the truest concept of the solar system, that of Aristarchus, went into eclipse and its author was accused of impiety in downgrading our planet by removing it from the centre of things. More serious still, Hipparchus, regarded by many as the greatest astronomer of antiquity, rejected on scientific grounds the idea of Aristarchus that the sun is central. The basis for the rejection was the inability of the Aristarchus model to account for the observed movements of the wanderers. We know today that it could not do so because the planets were pictured as moving in circles, instead of ellipses.

Thus this 600-year period of ferment and progress in astronomy came to an end with the concept of spheres within spheres more deeply entrenched than ever. Its final great proponent was Ptolemy who lived in Alexandria in the second century A.D. His dogma, which remained virtually unchallenged for 1,400 years, was an elaboration of the concentric spheres described by Aristotle, but with the addition of 'epicycles'. These were comparatively small circles described by planetary bodies around points that, in turn, moved in larger circles. The earth sat majestic and motionless at the centre of everything. However, some of the circles were off-centre, the better to explain various movements. Ptolemy, like some of the early thinkers, regarded the spheres as mathematical concepts, rather than crystalline objects. His epicycles still proved inadequate and, as observing methods improved and records of past observations became more extensive, more and more epicycles were added to overcome the difficulties. By the sixteenth

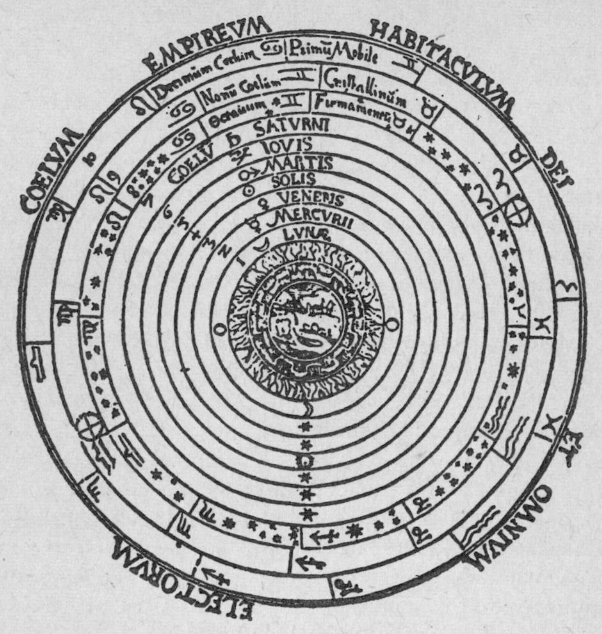

The universe, as it was envisioned until the Renaissance, is shown in this illustration from Peter Apian's *Cosmographia*, published in 1539. The earth is at the centre and the sun moves in a circle between the circles of Venus and Mars. Beyond the star-studded shell lies the Empirean heaven, habitation of God and all the Elect.

century more than eighty spheres were being called upon to account for the observed motions.

Astronomy became, in fact, a most disheartening activity. It has been said that not one discovery of significance was made from the death of Ptolemy until the time of Copernicus, fourteen centuries later. The fundamental concepts were not to be challenged. A scholarly work was looked upon with suspicion if it did not pay homage, in particular, to the ideas of Aristotle. But this was a state of affairs that could not go on for ever. It came to a dramatic end in the Renaissance, when men's minds were suddenly opened to the vastness of the universe and the idea that there must be many inhabited worlds was revived.

3

Science Reborn

WHEN we think of the overthrow of the earth-centred view of the universe that occurred during the Renaissance, the name of Nicolaus Copernicus comes immediately to mind, although the great Polish astronomer who originated what we call the Copernican system clung to many of the old ideas.

Copernicus not only shifted the centre of things from the earth to the sun, but, in the minds of men, he set the world spinning. At long last it began to become accepted that the celestial sphere stands still; that it is our earth that turns. To Copernicus, however, it was necessary to retain the epicycles – the movement of planets in small circles around points that travel in large circles – although his proposals halved the number of circles needed to account, in a gross way, for observed motions within the solar system. In the words of Herbert Dingle, Professor of History and the Philosophy of Science at University College, London, 'He clung as firmly as the most orthodox medieval philosopher to the machinery of spheres and to the Aristotelian principles of perfect celestial substances and uniform circular motions.'

Copernicus's work on the solar system was published as he lay dying, in 1543. Less than a century later, on 7 January 1610, Galileo Galilei, having brought the telescope to bear on the heavens, discovered four of the moons of Jupiter, showing that the orbiting of small bodies around larger ones may be a widespread phenomenon. Yet even Galileo still clung to the epicycles. Meanwhile the Danish astronomer Tycho Brahe demolished the still widely held idea that concentric crystal spheres enclose the earth. He showed that the paths of comets pass unobstructed through these hypothetical shells, but he refused to accept the Copernican model of the solar system, either on religious grounds or because it seemed inadequate to explain the observed motions. He said the sun and moon circle the earth and the planets circle the sun, much as did Heraclides almost 2,000 years earlier. However, the painstaking observations and calculations of the movements of Mars, made from Brahe's observatory near Prague, paved the way for his assistant, Johannes Kepler, to recognize the basic laws of orbital movement.

Kepler, like Copernicus, had one foot through the door to modern science and one foot still in the Middle Ages. He was unable to divorce astrology, prophecy and mysticism from his astronomy. His studies were interrupted when his mother was put on trial as a witch. His interest, and that of his contemporaries, in Mars was rooted in the strangeness of its orbit, for, among the planets, its movement most stubbornly resisted description in terms of spheres and epicycles. Using Brahe's data, Kepler finally discovered why: its path is an ellipse with the sun at one of the two foci. When Kepler looked at the other planets he found that they, too, flew elliptical paths. Suddenly the need for that absurdly complex system of spheres and epicycles vanished; the solar system became beautifully simple. The fact that satellites travel in ellipses was but one of the laws discovered by Kepler. Another was that a line joining the centre of a planet with the centre of the sun sweeps out equal areas in equal periods of time. Thus when a planet is in the portion of its orbit nearest the sun, it must move faster than it does when it is most distant in order to sweep out the same area in the same time.

However, having stated his laws, Kepler could not explain them. That remained for the mathematical genius of Sir Isaac Newton, almost a century later. Newton's laws of gravitation, as demonstrated in the laboratory, accounted for planetary motions. This had profound implications, for it suggested that nature behaves the same everywhere. Those cold pinpoints of light that we call stars were so distant and seemingly unrelated to things on earth that they had been considered residents of another realm, governed by different physical laws.

So completely do we today take for granted the universality of nature's laws that it is difficult to appreciate the extent of this revolution in thought. Modern astronomy is founded on the premise that, for example, the spectrum of hot sodium in the laboratory is identical to the spectrum of hot sodium within the most distant star. Yet until the Renaissance the view was deeply entrenched that there are two realms: the earth with its laws and manifestations, and the heavens with completely unrelated laws and phenomena. Integration of the two into a single structure made life elsewhere seem far more reasonable.

Newton's theory of gravity also suggested to him a mechanism for star formation that was prophetic:

> ... if the matter was evenly disposed, throughout an infinite space [he wrote] ... some of it would convene into one mass and some into another, so as to make an infinite number of great masses, scattered at great distances from one to another throughout all that infinite space. And thus might the sun and fixed stars be formed ...

The difficulty with Newton's idea that stars are formed by 'falling together' was the problem of ignition. What starts them 'burning'? It was only with the recent discovery that great pressure can produce thermonuclear reactions that this question was answered,

The concept of a universe populated by countless suns like our own, as expressed by Newton in the above quotation, was an essential ingredient in restoring the idea that there must be other inhabited worlds, but it was also necessary to shatter, in men's minds, the star-studded sphere that had enclosed the universe since before the time of Plato. The one generally credited with this was Thomas Digges, an English astronomer and mathematician of the sixteenth century who probably did more than any other man to promote acceptance of the Copernican model in Britain. More important, he realized that, with the stars stationary instead of circling the earth daily, they no longer had to be wedded to a sphere. From this Digges concluded that the stars were distributed through space, the dimmer ones being farther away. The heavens, he said, are 'garnished with perpetvall shininge gloriovs lightes innvmerable, farr excellinge ovr sonne both in qvantitye and qvalitye. ...' Furthermore, he said, the greatest part of the stars 'rest by reason of their wonderfull distance inuisable vnto vs.'

In spite of the majestic breadth that he attributed to the universe, Digges saw the sun as its centre. He said that the sun, like

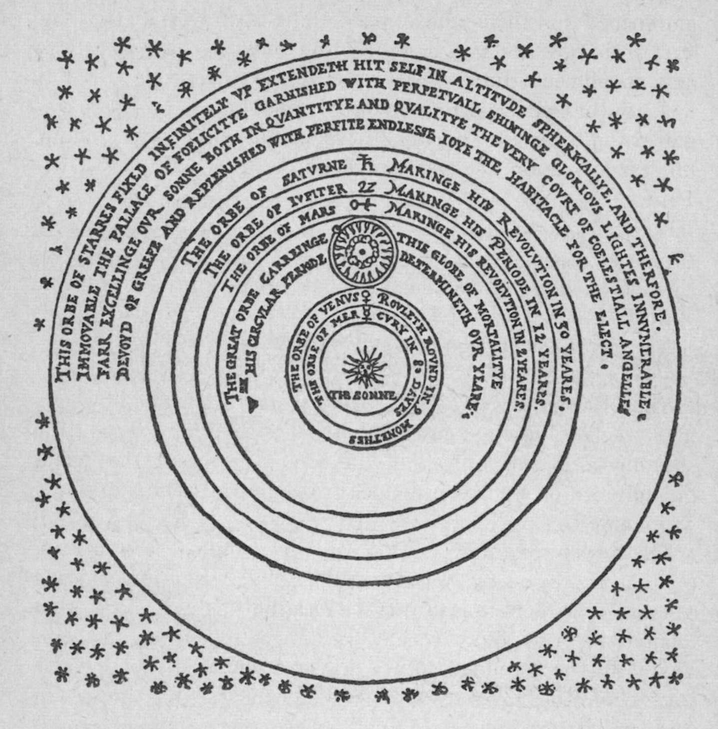

In the universe depicted by Thomas Digges in 1576, only a generation after Peter Apian, the sun is at the centre. The moon circles the earth and the stars are distributed uniformly through space beyond the solar system.

a king 'in the miidest of al raineth and geeveth lawes of motion to ye rest.'

The views expressed by Digges were carried even further, in the time of Newton, by the Dutch mathematician and physicist Christiaan Huygens. He stated that the sun is nothing more than just another star and then asked: '... why may not every one of these stars or suns have as great a retinue as our sun, of planets, with their moons, to wait upon them?' The qualities, he added, that we find on the planets of our own solar system 'we must likewise grant to all those planets that surround that prodigious number of suns. They must have their plants and animals, nay and their rational creatures too, and those as great admirers, and as diligent observers of the heavens as ourselves . . .'

Present-day champions of extraterrestrial life could hardly have stated the case better than Huygens, 300 years ago. His scientific credentials were of the highest. Not only was he the discoverer of Saturn's rings; he was the first to use pendulums in clocks and he developed a wave theory of light that was of historic importance.

Huygens anticipated one objection to his proposal: if there are planets orbiting other stars, why can we not see them? His reply was that the stars are too far away. He pointed out that they do not seem to come any nearer, or draw any farther away, in the course of the earth's annual journey around the sun, indicating that they are almost incredibly far away. Furthermore, he said, people are misled by the seeming size of the brighter stars. Thus the brightest star of the heavens, Sirius or the 'Dog Star', seems to the naked eye to be as big in the sky as the planet Jupiter. When one looks at Jupiter through a telescope it can be seen as a sphere of considerable size. However, no matter how big the telescope, Sirius is still a shapeless blob of light that would appear as a pinpoint, were it not for a number of blurring effects, including turbulence in the atmosphere, the response of the eye's retina and even the nature of light itself.

To illustrate his point, Huygens cited 'the nature of fire and flame which may be seen at such distances, and at such small angles as all other bodies would actually disappear under'. This, he added, was a thing 'we need go no further than the lamps set along the streets to prove.'

It was this illusion of nearness, said Huygens, that led Brahe to doubt that the earth circles the sun, for he thought that the stars would then vary in brightness. Likewise Kepler was deceived into believing that the stars are closely packed together. If they were as distant from one another as the sun is from the nearest stars, he reasoned, only a few would be visible.

Huygens was not the first man of that period to startle his contemporaries with talk of other worlds and their inhabitants. Roughly a century before he set forth his reasoning in his *Cosmotheros*, a young Dominican monk, Giordano Bruno, was denounced as a heretic to the Inquisition in Naples and fled to become a roving scholar, teaching at the great intellectual centres of Europe. He visited Elizabethan England and, because of resemblances between his poems and certain Shakespearean sonnets and plays, some believe he knew William Shakespeare. He may also have known Digges, proponent of a boundless universe.

Bruno preached the Copernican model of the solar system with missionary zeal and went on to argue for an infinite universe with infinite worlds, his approach being more philosophical than that of Digges and Huygens. The 'fixed' stars are not fixed at all, he argued,

> For if we could observe the motion of each one of them, we should find that no two stars ever hold the same course at the same speed; it is but their great distance from us which preventeth us from detecting the variations. . . . There are then innumerable suns, and an infinite number of earths revolve around those suns, just as the seven we can observe revolve around this sun which is close to us.

Anticipating Huygens by a century, he argued that these planets cannot be seen because the stars are far too distant. He likewise predicted that there may be additional planets in our own solar system, invisible either because of their distance, their small size or the poor light-reflecting properties of their surfaces (since then three additional planets and a great number of smaller asteroids have, in fact, been discovered).

If one accepted the view that the universe is infinite, which he felt was unavoidable, then its being peopled by a limited, and therefore 'imperfect', population of intelligent beings was to Bruno incompatible with the infinite goodness or perfection attributed to God and His works. Thus, he said, 'infinite perfection is far better presented in innumerable individuals than in those which are numbered and finite.' He therefore concluded that there must be an infinite number of morally imperfect beings, inhabiting the infinitude of worlds.

Bruno's ideas shocked his contemporaries. The Copernican view was far from being accepted. In fact in 1616, sixteen years after Bruno's death, Rome declared the Copernican system dangerous to the faith, an action which led to Galileo's trial and humiliating recantation before the Inquisition.

Giordano Bruno has been described as a firebrand, a romantic, and as architect of 'the greatest philosophical thought structure of the Renaissance'. The concept of an infinitude of worlds and rational beings gave flight to his soul. He did not trim his sails to avoid making enemies. Despite the fact that he was sought by the Inquisition, in 1591 he boldly returned to Venice, which had been acting somewhat independently of Rome in such matters. Not long after his arrival he was seized and subjected to prolonged interrogation and trial. Among the charges against him was that he had praised Queen Elizabeth of England, a heretic and enemy of the Church. He replied that his praise was for her fine qualities, not for her opposition to Rome. He was subjected to what today would be called brainwashing and at one point, broken and abject, he confessed to a variety of offences, only to relapse later to his unorthodox position – much like Savonarola and Joan of Arc.

In 1600, after six months of imprisonment in Rome, he was burned at the stake. He is said to have been defiant to the end. 'I have fought, that is much,' he said; 'victory is in the hands of fate.' As he was dying in the flames someone thrust a crucifix into his hands, but he fiercely pushed it away.

The lines of reasoning developed by Bruno, Huygens and the like ignited the minds of thinking men throughout Europe. In Milton's *Paradise Lost*, written in the mid-seventeenth century during Huygens's lifetime, the angel Raphael tantalizes Adam with the thought that the moon and planets in other solar systems may be inhabited. If there be land on the moon, he wrote, might there not be

> Fields and inhabitants? Her spots thou seest
> As clouds, and clouds may rain, and rain produce
> Fruits in her softened soil, for some to eat
> Allotted there; and other Suns, perhaps,
> With their attendant Moons, thou wilt descry,
> Communicating male and female light –
> Which two great sexes animate the World,
> Stor'd in each Orb perhaps with some that live.

That so vast a realm as the universe should be devoid of life, serving no other purpose than to convey tiny fragments of starlight to the earth, was, the angel said, 'obvious to dispute'. But he urged Adam to leave such questions to God and, instead, to rejoice in his paradise and his fair Eve:

> Dream not of other worlds, what creatures there
> Live, in what state, condition or degree . . .

Milton's belief that other worlds may exist, but that we can never observe them, was reflected seventy years later in Alexander Pope's *An Essay on Man.* However, Pope, like those today who have thought about this when direct knowledge of life beyond the earth is conceivable, saw in such elusive knowledge a means of understanding ourselves:

> Of man, what see we but his station here,
> From which to reason, or to which refer?
> Through worlds unnumber'd though the God be known,
> 'Tis ours to trace him only in our own.
> He, who through vast immensity can pierce,
> See worlds on worlds compose one universe,
> Observe how system into system runs,
> What other planets circle other suns,
> What varied Being peoples ev'ry star,
> May tell why Heav'n has made us as we are.

The despair of Milton and Pope at ever observing life on other celestial bodies was not shared by all their contemporaries. In 1638 Bishop John Wilkins, brother-in-law of Oliver Cromwell, published a work (originally anonymous) entitled: *The Discovery of a World in the Moone, or a Discourse tending to prove that 'tis probable there may be another Habitable World in the Planet.* In a later edition he added a *Discourse concerning the Possibility of a Passage thither*, in which he discussed the journey in terms of air flight.

Wilkins was one of those men of that period who bristled with ideas. He proposed the use of submarines (as yet uninvented) for voyages under the polar ice, and helped organize weekly meetings of savants to explore the exciting avenues of scientific speculation opened to them by what they called the New Philosophy. Those who attended the talks, a number of them held in Wilkins's own quarters, included such men as Robert Boyle, famed for his law regarding the compression of gases, Sir Christopher Wren, the architect, and Samuel Pepys, the diarist. The group was finally chartered and became known as the Royal Society, one of the most distinguished associations of scientists ever formed, Newton being an early member.

This was a period of colonial expansion and Wilkins proudly predicted that the first flag to fly on the moon would be British – in contradiction to the claim of another imaginative scientist, Johannes Kepler, that the first flag there would be Germanic. Nor was Wilkins the only man of that time to propose going to the moon as a way to settle the question of extraterrestrial life. On the Continent a counsellor to the French court, Pierre Borel, wrote that 'the way whereby one can learn the pure truth concerning the plurality of worlds' is by aerial navigation (what we would call space travel). One of Borel's contemporaries went even further and told of his own voyage to the moon. This pioneer in science fiction was Cyrano de Bergerac, known today chiefly as the hero of Rostand's play about the swaggering, skilful, romantic swordsman who fought all comers to defend the honour of his elongated nose. Cyrano was a real-life figure who, after being wounded in a war with the Spanish, turned to writing satirical novels. In one of them he meets 'a little man, entirely naked', who speaks to him in an utterly strange tongue and proves to be a visitor from another planet. Cyrano himself visits the moon and, for a time, is caged by one of its unscrupulous inhabitants as a freak. He is then charged with impiety for claiming that the 'moon' from which he came is a world, whereas the 'world' at which he had arrived is a moon. To save himself from death, he goes about among the moon people, assuring them that their moon is a world because 'that is what the Council finds it proper that you believe.'

Far more effective in stimulating serious thought about the possibility of far-flung life was a book, written a generation later, by a French scientist-satirist whom Voltaire called 'the most universal genius that the age of Louis XIV has produced'. His name was Bernard le Bovier de Fontenelle, nephew of the playwright Corneille, and in some respects Voltaire's predecessor as the chief gadfly of the French social scene. He helped write an opera when barely twenty and later became official historian of the French Academy of Sciences. However, the book that brought him fame throughout Europe was *Conversations on the Plurality of Worlds*, published in 1686 as a series of talks with a fictitious marchioness.

His argument was that other bodies of the solar system are inhabited, although those who live on Mercury, nearest planet to the sun, 'are so full of fire, that they are absolutely mad: I fancy they have not any memory at all.' On the other hand, he said, the residents of Saturn, outermost of the known planets, 'live very miserably . . . the sun seems to them but a little pale star, whose light and heat cannot but be very weak at so great a distance. They say Greenland is a perfect bagnio, in comparison to this planet. . . .' He proposed that visitors from other worlds travel by comet.

The marchioness replied, at one point, with a comment applicable to some of today's speculations: 'You know all is very well,' she said, 'without knowing how it is so; which is a great deal of ignorance, founded upon a very little knowledge.'

Although this fantasy, half-serious in its effort to set people's minds into new trains of thought, appeared almost a century after Giordano Bruno died at the stake, such ideas were still considered dangerous. In particular, Fontenelle's satire on relations between Rome (Catholicism) and Geneva (Calvinism) led to his denouncement as an atheist by Tellier, confessor to Louis XIV. According to Voltaire, Fontenelle was saved only through the intervention of his friend Marc-René de Paulmi, marquis of Argenson, who was then Lieutenant of the Police.

Fontenelle's concept of many worlds, expressed in so readable and popular a manner, excited people far and wide. Mikhail Vasilievich Lomonosov, the so-called father of Russian science, read the book in its French original and was deeply impressed. His belief in many inhabited worlds was reflected in a number of the poems – scientific and satirical – written by this versatile man and, when the first translation of Fontenelle's book was suppressed by the Russian Church, Lomonosov saw to it that a second translation into the Russion was published. It might even be argued that seeds thus planted in eighteenth-century Russia bore fruit in the work of Otto Struve after he came to America and presided over mankind's first serious attempt to reach out and make contact with intelligent beings elsewhere.

Nevertheless, the flights of fancy that flowered in the uninhibited minds of many eighteenth-century thinkers came square up against the intellectual discipline of the New Philosophy, which demanded that every hypothesis be subjected to the test of experiment. The curse of scientific thinking prior to the Renaissance had been the adherence to dogma, which led to perpetual doctoring of inadequate theories – as with the epicycles – rather than to the testing of all concepts, no matter how well-established, in the light of new discoveries. However, there was no means at hand to test the idea that there are intelligent beings on distant planets. In fact the very existence of planets beyond the solar system could not be demonstrated.

Furthermore, a succession of serious difficulties arose in the path of those arguing for such life. The more that was learned of the moon and planets of our own solar system, the less hospitable they appeared. The moon proved to be totally dry and airless. Mercury was far too hot and the outer planets far too cold. Apart from the earth itself, this left only Mars and Venus. Then the argument that other stars must have planets, like our sun, was dealt what seemed a fatal blow with the general acceptance, early in this century, of a new theory of the origin of the solar system. It was proposed that the planets were formed when the gravitational attraction of a passing star sucked out portions of the sun, which then cooled externally to form bodies with solid crusts.

So vast are the distances between the stars that the chances of frequent near collisions, such as the one depicted in this hypothesis, seemed negligible. As noted by the Russian scientist A. I. Oparin, if the sun were the size of an apple, on the scale of the universe the nearest star would be as far away as the distance from New York to Moscow. Hence the solar system appeared to be a freak and life on earth came to be regarded as unique. Discussion of life elsewhere was relegated to the authors of science fiction. Only with the step-by-step dissolution of the difficulties has serious discussion of the subject been resumed, much as it was after the Renaissance, and it has even become possible, within limits, to test the hypothesis in the spirit of the New Philosophy in experiments such as that of Project Ozma.

The accumulation of data that has provided meat for these discussions falls into three main categories, forming the subject matter of the next nine chapters:

1. Evidence for a boundless, or almost boundless, extent to the universe and for an almost limitless number of stars.

2. New theories for formation of the solar system, suggesting that it is a common phenomenon, plus evidence hinting at the presence of other such systems.

3. Development of biochemistry to the point where some believe life will inevitably arise on any planet whose environment is similar to that of the earth, as well as evidence that at least some steps toward the evolution of life have occurred elsewhere.

4

Is Our Universe Unique?

IN recent decades science has opened a series of huge eyes upon the cosmos. The mirrors of the three great California telescopes, from 100 to 200 inches in diameter, can collect light from a patch of sky no larger than that covered by a pea at arm's length. Although these instruments weigh many tons, their delicate clockwork can keep them trained on the same spot of the moving heavens for hours, while a photographic plate soaks up light coming from vast distances. When developed, such a plate is peppered with stars – a small sample of the billions in our galaxy. More awesome still: for each of these stars, in many photographs, there can be seen a distant galaxy beyond.

Yet information on our own galaxy is surprisingly meagre, for even the 200-inch telescope on Mount Palomar cannot see its core. The latter is completely hidden by clouds of dust and gas that also conceal other regions of the galaxy. Only with the new radio telescopes is it possible to 'see' the galactic core by means of radio emissions generated from the fearsome processes at work there.

The observation and reasoning that have finally relegated our sun to the status of an average star among billions (in an average galaxy among billions) began when Galileo turned his crude telescope on the Milky Way and saw that, at least in part, it was made up of myriad stars. Some proposed that the Milky Way constitutes a ring of stars around the universe, but in 1750 Thomas Wright of Durham, England, private tutor and instrument maker, published *An Original Theory or New Hypothesis of the Universe*, in which he said the stars are scattered more or less uniformly between two parallel planes, as though in a gigantic sandwich of unknown extent. The sun, he said, is midway between the two planes so that, when we look 'out' towards either of the planes, we see a minimum number of stars, whereas if we look parallel to the planes we see the seemingly impenetrable star clouds of the Milky Way. While the sun is on or near the centre plane, it is not necessarily at the centre of the whole system, he said. It is, he argued, 'not only very possible, but highly probable ... that there is as great a multiplicity of worlds, variously dispersed in different parts of the universe, as there are variegated objects in this we live upon. Now, as we have no reason to suppose that the nature of our Sun is different from that of the rest of the Stars ... how can we, with any show of reason, imagine him to be the general centre of the whole... ?'

Despite this reasoning astronomers continued, well into this century, to assume that our sun is at or near the centre of things, if only because the Milky Way seems to surround the solar system so uniformly.

Wright initiated another controversy resolved only in recent decades. He proposed that our Milky Way (or 'galaxy', based on the Greek word for milk) is but one of many such systems. There are, he said, 'a Plenum of Creations not unlike the known universe'. This, he added, 'is in some degree made evident by the many cloudy spots, just perceivable by us, as far without our starry regions in which ... no one star or particular constituent can possibly be distinguished; those in all likelihood may be external creations, bordering upon the known one, too remote for even our telescopes to reach.'

It was the great German metaphysician Immanuel Kant who seized upon this idea and made it famous in his discussion of 'island universes' published only five years later, in 1755. He pointed out that, if these universes (galaxies) were disc-shaped they would, when viewed at an oblique angle, appear elliptical. This was, in fact, the shape of many of the nebulous objects visible in telescopes of that time.

The first systematic study of our galaxy's structure was carried out by William Herschel, astronomer to George III of England, by counting stars in different parts of the sky and then seeing what model of the universe best suited his observations – a technique elaborated upon early in this century by the Dutch astronomer, Jacobus Cornelis Kapteyn. In 1785 Herschel announced that the sun is near the centre of a flat system of stars roughly five times wider than it is thick. He agreed at first that the nebulae, or cloudy spots, referred to by Wright are distant galaxies of the same sort. However, when his improved telescope failed to show individual stars in many of these luminous objects, he revised his view and said they were strange bodies of 'shining fluid'.

For more than 100 years thereafter astronomers wavered back and forth. By the middle of the nineteenth century there were telescopes of sufficient power to resolve stars in some spiral nebulae, but at the same time the emergence of spectroscopy – the science of light analysis – opened a new era in astronomy and raised doubts with regard to the nebulae. Because the stars are too far away to be seen except as points of light, most of present-day knowledge concerning them is based on study of their spectra. From this we have learned not only their composition, but their surface temperatures, magnetic properties, rates of rotation, rates of movement toward or away from us and duality (in some binary systems). It was found that the light from some irregularly shaped nebulae showed a 'bright-line' spectrum typical of luminous gases, rather than of stars, and this led many to doubt that any of the nebulae are, in fact, distant star clouds. Furthermore, the more numerous, symmetrically formed nebulae seemed to be concentrated toward the poles of our galaxy, above and below its central plane, as though forming an integral part of it.

Today we know that there are two broad categories of nebulae: those that lie within the galaxy and, in many cases, are irregular clouds of gas made luminous by radiation from nearby stars, and the symmetrical type that lie at great distance and are, in fact, other galaxies. We also know that they seem to be concentrated toward the poles of our galaxy simply because dust clouds dim our vision along the plane of the galaxy. It was not until 1924 that it was finally proved, by Edwin P. Hubble of Mount Wilson Observatory, that there are other galaxies and that our own is no more central in the universe than is our earth in the solar system. The tools with which he demonstrated the true nature of the distant galaxies were the same as those that enabled Harlow Shapley, while at Mount Wilson, to show that the sun is very far from the centre of our own galaxy. The tools were a remarkable class of stars known as Cepheid variables.

As early as the sixteenth century it was noticed that a star in the constellation of the Whale appears and vanishes at irregular intervals. Its period – that is the time between each reappearance – is roughly one year. Additional 'variable' stars were found and it was discovered that most of them do not vanish, like a light being switched on and off, but rather dim and brighten, often with a characteristic rhythm. With one class of variables it became evident that they are 'waltzing' with a dark companion. When the latter comes between the earth and the visible star, the light of the latter is dimmed or eclipsed. However, another type dims slowly, then rises to maximum brightness rapidly, with accompanying changes in spectrum that show the star to be actually pulsing.

One of the most prominent of these stars appears in the constellation Cepheus; hence they are known as Cepheid variables (a category that in recent years has been further subdivided). The nearest and best known is Polaris, the North Star, whose pulse rate is 3·97 days. The periods of the various kinds of variable stars range from minutes to more than a year.

Towards the end of the nineteenth century the Harvard College Observatory began a systematic search for variables by comparing photographs taken of the same region of the heavens at different times. Various methods were used to speed up the process, such as laying a positive plate over a negative plate of the same star pattern, exposed at a different time, thus cancelling out all stars whose brightness was unchanged. By 1910 some 4,000 variables were known, 3,000 of them discovered at Harvard. Data from the southern heavens were obtained at a field station operated by the university high among the Andes, at Arequipa, Peru. At the observatory in Massachusetts Henrietta S. Leavitt studied the variables appearing in photographs of the two Clouds of Magellan, taken from Arequipa. These two clouds of stars are now classed as irregular galaxies and are our own galaxy's nearest neighbours in the cosmos.

Miss Leavitt and her co-workers listed 1,777 variables in the two clouds, and she picked for study those in the Small Magellanic Cloud for which reliable data on pulse rates and luminosity were available. In 1904 she published preliminary findings on seventeen of them and eight years later, with firm data on eight more, she felt confident of her discovery. There is, she said, 'a remarkable relation between the brightness of these variables and the lengths of their periods'.

The slower the pulse rate of the star, the brighter its light, the rule being applicable both at maximum and minimum. She pointed out that the stars, clustered in a compact and extremely distant cloud, could all be considered the same distance away and therefore the differences in their brightness were intrinsic and not a result of differences in distance.

At first the discovery seemed of interest only in trying to explain why variable stars behave as they do. However, Danish astronomer Ejnar Hertzsprung almost immediately saw the significance of these stars as a potential yardstick for the universe. Since the dimming of light by distance follows a well-established law, once the distance to one Cepheid variable was known its intrinsic brightness could be calculated. Then, by simply timing the pulse rate of any other Cepheid, it would be possible to reckon its distance also. The trouble was that none were near enough so that their distance could be triangulated, even using the longest base line available to us: that joining opposite sides of the earth's orbit around the sun. However, a more crude method was available, namely that of analyzing rates of apparent movement against the stellar background. On the average, the faster a star changes position among the other stars the nearer it is presumed to be. In this way Hertzsprung, in 1913, obtained rough distances for the thirteen nearby Cepheids on which he had adequate data. Shapley then attacked the same problem and the yardstick, though still imperfect, was forged.

The principle of the Cepheid yardstick can be envisioned in terms of a night sky filled with lighthouses flashing at what seem random rates. Some lights are brilliant; some so dim they can only be seen through binoculars, but their brightness cannot be used as a measure of distance, since it is known that some lighthouses contain lights of blinding brilliance, whereas other have bulbs of only a few watts. Then it is found that the rate at which each light flashes is an indication of its intrinsic brightness. All that then has to be done is to find the distance to one lighthouse in order to determine how far it is to all the others.

Having acquired the yardstick, Shapley, then at Mount Wilson, immediately used it to attack a problem that had long perplexed astronomers, namely the peculiar distribution of globular clusters. Each of the latter resembles a beautifully symmetrical swarm of bees. They consist of many tens or hundreds of thousands of stars. The puzzling feature was their concentration toward the constellation Sagittarius. One-third of all those that can be seen lie there, within only two per cent of the celestial sphere, and most of the others are not far away.

Fortunately the clusters were extremely rich in various stars that seemed identical to those of the Magellanic Clouds. (The fact they actually differed did not affect the validity of Shapley's conclusions.) He set to work with his yardstick, charting in three dimensions the true distribution of the globular clusters. He found that, instead of being grouped close together, as they seemed, they are spread throughout a spherical region so large it was incredible to most of his contemporaries. Their distribution within this sphere proved to be remarkably symmetrical about the central plane of our galaxy. Of those that he studied, forty-six were above this plane and forty-seven were below it. Most startling of all, the centre of this sphere – which he rightly assumed to be the centre of the galaxy – lay, according to his calculations, some 50,000 light years away towards Sagittarius, one light year being the distance travelled by light in a year at 186,000 miles a second. By contrast, light takes only eight minutes to reach us from the sun. Shapley estimated that the disc-like system of stars forming our galaxy is symmetrically enclosed by the globular clusters and is 300,000 light years wide and 30,000 light years thick.

His findings, published in 1918, created a sensation. His model of the galaxy was ten times bigger than the accepted model and it seemed hardly believable that the sun could be so far from the galactic centre. At that time Shapley was, in his own words, 'a callow youth', and notable among his critics was Heber D. Curtis of the Lick Observatory, a big name in contemporary astronomy. So heated were the discussions and so widespread the interest that the National Academy of Sciences took the unusual step of staging a public debate between Shapley and Curtis. It took place on 26 April 1920 before a packed house at the academy headquarters in Washington. Subsequent work has shown that, in principle, Shapley was right. Like others of that period, he was unaware of the extent of dust and gas that dims the light from distant stars and hence overestimated his distances. Present figures for the width of the galaxy range from 80,000 to 100,000 light years and place the thickness at some 25,000 light years.

On another question that figured in what some call the 'Great Debate' at the academy – the perennial controversy over island universes – Curtis proved to be right and Shapley wrong. Shapley, on the basis of a report (later proved erroneous) that some of these objects (spiral nebulae) are drifting across the stellar background, said they are comparatively near, whereas Curtis said they are other galaxies like our own. Four years later the matter was settled once and for all when Edwin P. Hubble reported that, using the new 100-inch telescope on Mount Wilson, he had identified Cepheid variables in the Great Spiral Nebula of Andromeda, the only spiral star system near enough to be seen with the unaided eye. Cepheids, soon seen as well in other such objects, showed them to lie at distances of 1,000,000 light years or more. Clearly they were sister galaxies to our own, and by looking at their spiral star clouds surrounding a bulbous, bright-glowing core it was possible to form an idea of the nature of our galaxy. It has been said that it is about as difficult for us to chart the Milky Way Galaxy, from within it, as it would be for someone standing in Central Park to draw a map of New York City.

Shapley was deeply conscious of the significance of his displacement of the sun from near the centre of things. 'By this move,' he wrote in 1958, 'we have made a long forward step in cosmic adjustment – a step that is unquestionably irreversible. We must get used to the fact that we are peripheral, that we move along with our star, the sun, in the outer part of a galaxy that is one among billions of star-rich galaxies.'

Thus, by the late 1950s, we were coming back to the trains of thought that gripped men in the days of Giordano Bruno. 'No field of inquiry,' Shapley wrote after his retirement as head of the Harvard College Observatory, 'is more fascinating than a search for humanity, or something like humanity, in the mystery-filled happy lands beyond the barriers of interstellar space.'

He reasoned that, no matter how scarce such lands may be, the almost infinite number of stars demands that they exist.

As far as we can tell [he wrote] the same physical laws prevail everywhere. The same rules apply at the center of the Milky Way, in the remote galaxies, and among the stars of the solar neighborhood. In view of a common cosmic physics and chemistry, should we not also expect to find animals and plants everywhere? It seems completely reasonable; and soon we shall say that it seems inevitable.

5

The Solar System: Exception or Rule?

THE question of whether or not we are alone in the universe depends heavily on the manner in which our planet, and the solar system of which it is a part, were formed. Running through the long debate on this subject have been two rival theories, frequently subjected to scientific attack and modification, but rooted in concepts that, essentially, came out of the eighteenth century.

One, set forth in 1745 by the French naturalist George-Louis Leclerc, comte de Buffon, proposed that a comet, apparently viewed by him as comparable in size to a star, struck the sun and knocked off lumps that became the planets. Since the blow was off-centre, the sun was set to spinning and the planets not only were cast into orbits around the sun but spun on their own axes. In fact some spun so fast they threw off additional material that became moons. As noted earlier the distribution of stars in space is so thin that such a collision – or even a near miss – would be a freak event and the number of other solar systems in our galaxy would be negligible.

Ten years later Immanuel Kant proposed a quite different model. He envisaged a primordial universe consisting of gas that condensed into blobs of higher density. One mighty blob became the solar system. First, because of its spin, it flattened into a disc whose core, attracted to itself by gravity, 'fell' together to form the sun. Other cores formed nearby and became planets. In this manner, not one solar system, but innumerable ones, would be formed.

Kant's proposal was, to some extent, anticipated by some who came before him. More than a century earlier the great French philosopher and mathematician René Descartes suggested that the solar system formed from a cloud of primordial matter, although he made no reference to gravity to explain how this came about. Indeed, his theory was formulated before Newton recognized the laws of gravity. The natural and typical behaviour of all matter, Descartes said, was to move in a vortex, like water on the surface of a whirlpool—a theory reminiscent of the one devised by Democritus 2,000 years earlier. In the beginning, said Descartes, the stuff of the solar system moved in a series of vortices that rubbed against one another, dislodging material that drifted to the centre of each vortex. The sun thus formed in the middle of the large, central vortex and the planets and moon formed within secondary vortices.

In Kant's infancy the Swedish scientist and mystic Emanuel Swedenborg had begun to formulate a concept that had many points in common with that of Kant, but like many other of Swedenborg's advanced scientific ideas, it received little notice at that time. On the other hand, a few years later, the French mathematician Pierre Simon, marquis de Laplace, produced another closely related theory that remained a favourite in the scientific world for many generations. His starting point, like Kant's, was a nebulous cloud rotating around the sun. However, Laplace regarded the nebula as part of the sun itself. As this nebula contracted, its spin rate, because of a fundamental law of physics, had to increase. Laplace said this caused the cloud to flatten into a disc. As this in turn contracted, elements of it began circling the sun so fast that they remained behind in orbit, forming a succession of rings. It was from these rings, he said, that the planets condensed.

By then two of Kant's countrymen, Johann Daniel Titius and Johann Elert Bode, had observed a striking pattern in the spacing of the planets that seemed to offer a clue as to how they were formed, if only one could properly interpret its meaning. The pattern, known as Bode's Law, can be stated thus: the distances between planet orbits double for each step outward from the sun. More specifically, if the distance from the sun of the innermost planet, Mercury, is given as 4, the distances of the others are: $4 + 3$, $4 + 6$, $4 + 12$, $4 + 24$, et cetera. The solar system does not conform precisely to this 'law', but it can also be applied in a broad sense to the spacing of moon systems around the larger planets, such as Saturn (which has nine satellites). One planet in the Bode sequence is missing. It should lie between the orbits of Mars and Jupiter. In its place is the belt of asteroids that presumably are either the debris of a former planet or are material that never completed its coagulation into a planet.

Bode went further than the law that bears his name. Like many of his contemporaries, he speculated about the inhabitants of other planets and tried to apply to them a counterpart of his spacing principle. The material of which the planets are made becomes lighter at greater distances from the sun and Bode proposed a similar 'law' for their residents, namely that the populations became progressively lighter and more spiritual as one moves outward from the sun. This same idea was expressed at this time by Kant, who said the inhabitants of Venus and Mercury are too underdeveloped morally to be held accountable for their own acts, whereas those of Jupiter live in a superior state of happiness and perfection.

Swedenborg had a somewhat different view. In a series of dreams that he regarded (as do his followers today) as divine revelations, he was, he said, visited by spirits from other planets who described to him countless inhabited worlds. He was told that there were two races living on Venus, one gentle and humane, the other savage and cruel. Those inhabiting Mars, on the other hand, were said to be the finest residents of the solar system, resembling in piety the early Christians.

The more solid discovery of Bode and Titius, regarding the spacing of planets, harkened back to the idea of the ancient Greeks that the arrangement of the heavenly bodies was related to the spacing of notes in a musical chord. Even Kepler, discoverer of the laws of orbital motion, with his strange admixture of mysticism and mathematics, adhered to a somewhat similar view.

A further rule applicable to the solar system was recognized in the middle of the nineteenth century by E. Roche of Montpellier in the south of France, who sought to explain the rings of Saturn. He calculated the tidal stresses produced by the gravity of a parent body in objects orbiting it at various distances. For example, the gravity of the sun and moon are enough to produce our ocean tides and also to contort, slightly, the shape of the earth itself. Because of these earth tides all of Manhattan Island rises and falls a few inches twice a day, in terms of its distance from the earth's centre, but the gravity of our planet is sufficient to keep such stresses from tearing it apart.

Roche calculated that a satellite of the same density as its parent body could not resist these tidal stresses in an orbit closer than 2·44 times the radius of its parent. The innermost moon of Saturn lies outside this 'Roche Limit', but the outside edge of its rings is at 2·30 radii. Hence, Roche said, the rings consist of material that is too close to Saturn to coalesce into a moon. The innermost planet, Mercury, lies well outside the Roche Limit of the solar system.

In addition to these rules, there are others that should serve as clues to the manner in which the sun and planets were formed. In the first place all planet orbits are nearly circular; they lie roughly in the plane of the sun's equator (except the outermost, Pluto, which may once have been a moon of Neptune); and they all circle (and most of them spin) in the same direction as the sun's own spin. Secondly, the satellites of the sun fall into two general categories: the inner planets (including the earth), which are small, heavy, spin slowly and have few moons; and the outer planets, which are big, light, fast-spinning and escorted by a goodly company of moons.

For many years Buffon's collision hypothesis was forgotten and the cloud-condensation theories of Kant and Laplace dominated efforts to account for these rules. However, in the second half of the nineteenth century the Kant-Laplace concepts ran into trouble. James Clerk Maxwell of Scotland, who founded the modern theory of light, was also interested in the principle that keeps the rings of Saturn from condensing into moons. He calculated that, if all the material of the planets was spread uniformly through the solar system as a nebula, there would not be enough gravitational attraction between the particles—by a long shot—to draw them together and hold them there in face of the dismembering influence of the sun's own gravity. He persuaded the scientific world that planets could not condense from a nebula, as proposed by Kant, or from the rings of Laplace, and, indeed, this question of how planets could condense so near the sun has continued to plague theorists ever since.

Another difficulty with the Kant and Laplace theories arose

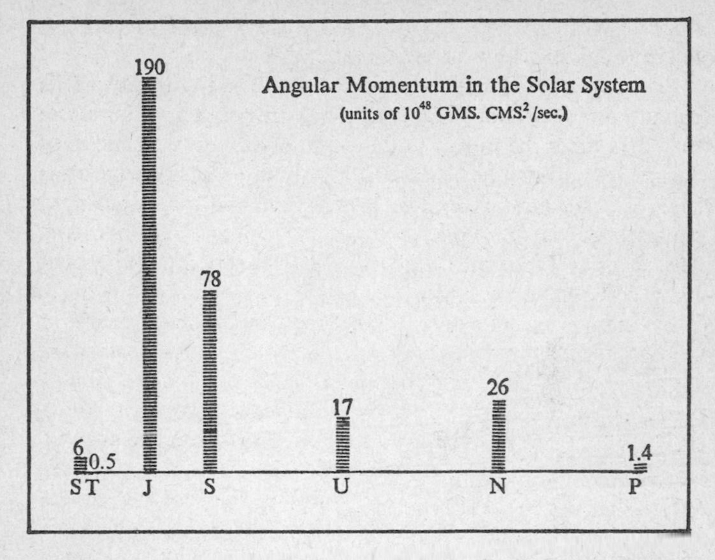

This diagram shows that within the solar system Jupiter, 'J', carries by far the largest proportion of the angular momentum – much more than the sun. The four inner, or 'terrestrial' planets carry so little that their combined angular momentum is shown as 'T'. The sun, 'S', is at the left.

from the peculiar distribution of angular momentum in the solar system. Angular momentum is that which keeps a flywheel spinning. Although the major planets, together, constitute less than 1/700th of the total mass of the solar system, they carry 98 per cent of its angular momentum. The sun, with almost all of the mass in the system, spins so slowly (about once a month) that it carries only 2 per cent of this momentum. If the sun and planets all condensed out of the same primeval cloud, the sun should spin once every ten hours or less.

Hence, early in this century there was a return to the collision idea of Buffon, but with attempts to spell out the sequence of events in greater detail and show how the sun's encounter with another star could have imparted much angular momentum to the resulting planets without doing so to the sun. Some said a star came close enough to cause great tides on the sun, tearing loose large amounts of material. Others reasoned that there must have been a grazing collision. T. C. Chamberlin and F. R. Moulton at the University of Chicago said the cataclysm must have produced eruptions of solar material that then assembled into planets. They identified the spiral nebulae seen in telescopes as other solar systems being formed, for the true nature of these objects as other galaxies was not yet accepted. Sir James Jeans and Sir Harold Jeffreys in England suggested that the encounter drew out a cigar-shaped filament of solar material whose fat, central section produced the largest planets – Jupiter and Saturn – and whose ends became the inner and outermost planets.

Such ideas linger in the older textbooks. They were welcome to those who took pride in the thought that we are unique. However, objections were raised against these theories, as they had been against those of Kant and Laplace. In 1939 Lyman Spitzer, Jr, of Princeton University pointed out that hot gas sucked from the sun would not condense into planets but would spread around the sun as a thin cloud. Other freak events were suggested as the starting point. Henry Norris Russell, also at Princeton, whose classification of the stars became a pillar of modern astronomy, proposed that, to begin with, ours was a binary system: two stars orbiting each other like waltzers on a dance floor. One of these stars was then destroyed in a close encounter with a third star, and the debris formed into planets with an excess of angular momentum.

Fred Hoyle, in England, likewise used a binary system as a starting point, but he proposed that one of the two stars blew up as a supernova. The latter can be likened to a nuclear explosion and sometimes makes a star brighter than an entire galaxy. Such cataclysms have been observed in our own galaxy only a half dozen times in the past 2,000 years.

Despite the distractions of World War II, scientists in both Russia and Germany pondered the difficulties raised by the collision idea. For example, it was argued that, if the earth was a piece of the sun, it could not have a core made largely of iron, in view of the scarcity of that metal in the sun. Yet there was ample evidence that the earth does, in fact, have a heart of iron. Arnold Eucken in Göttingen pointed out that, in a cloud of cooling solar gas, the first material to condense would be the iron. The earth, in his view, was thus born in a fearsome rain of molten iron that formed the core. The lighter components of the ball of hot gas were largely swept away.

However, in 1943 Eucken's countryman C. F. von Weizsäcker and the Russian mathematician and polar explorer Otto J. Schmidt returned to the idea that we were, somehow, born from a cloud of dust and gas. Von Weizsäcker proposed that eddies, or vortices, were formed symmetrically in a primordial cloud that was circling the sun. He worked out a pattern whereby these vortices would condense into the system of planets that exists today.

Schmidt said the planets formed from material captured by the sun from the clouds of dust and gas that we can see as shadow areas against the back-drop of stars. The spacing of planets, he said, was the natural result of competition for material scattered rather uniformly through space. He founded a Soviet school of thought that sought to explain the varying densities of the planets in terms of the sun's heat. The latter determined the weight of substances that would condense at any given distance from the sun, causing the inner planets to be heavy and the outer planets to contain both light and heavy materials.

In Holland, H. P. Berlage, Jr, revived the idea, advanced by Laplace almost two centuries earlier, that rings formed spontaneously, in the original nebula. However, all of these theories were criticized on the ground that the particles would not stick together, and therefore the accretion of planets could not begin. Gerard P. Kuiper, a Dutch-born astronomer who emigrated to the United States and became director of the Yerkes Observatory, tried to circumvent the difficulty. He said the vortices suggested by von Weizsäcker were too uniform and that the more probable cause was turbulence, here and there, where the material would become sufficiently dense to fall together, despite solar gravity.

One of the most active theorists on this subject has been Harold C. Urey, whose discovery of heavy water (deuterium oxide) won him a Nobel Prize. Beginning in the 1950s, while he was at the University of Chicago, he evolved a concept in which objects

THE ORIGIN OF THE SOLAR SYSTEM – THREE CONCEPTS

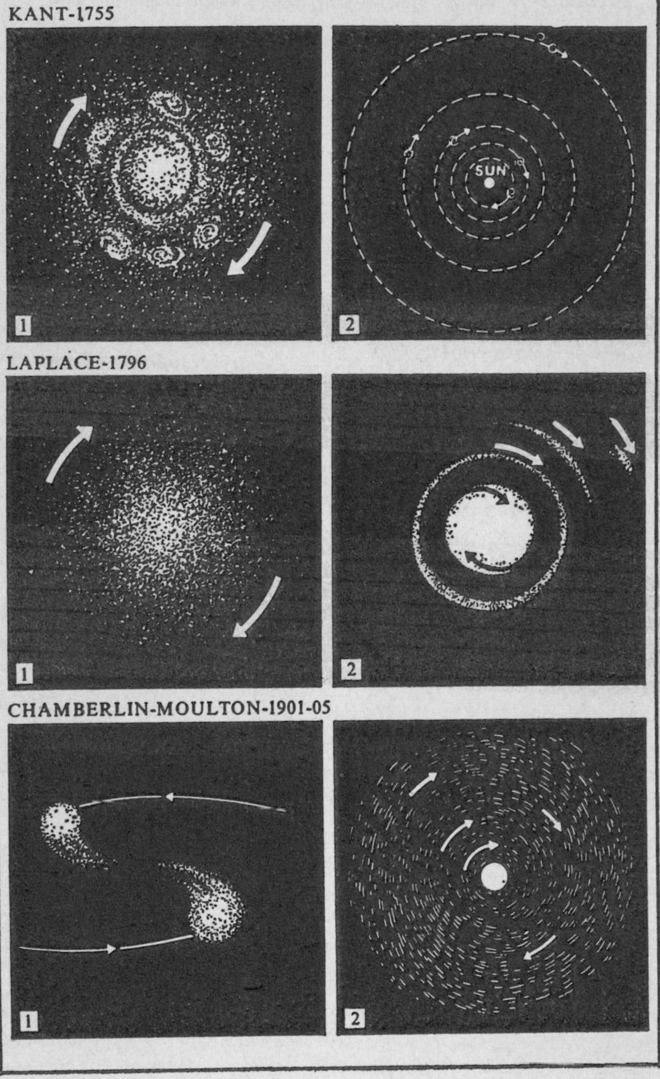

KANT-1755

1

2

SUN

LAPLACE-1796

1

2

CHAMBERLIN-MOULTON-1901-05

1

2

roughly the size of the moon formed from the dust and gas of the nebula. With the solar system full of such bodies, collisions were bound to occur, and the region became cluttered with debris similar in composition to the meteorites.

Meanwhile the pressure of light and outflowing gas from the sun swept lightweight material from the inner solar system. Hence, when the debris fell together again to form planets, the material was much heavier than that which condensed into the outer planets. The moon, according to Dr Urey's hypothesis, is a lonely remnant of the earlier stage, which would account for its lightweight composition.

Urey dismissed the old idea that the earth was formed from hot solar gas on the grounds that volatile substances, such as arsenic, are reasonably plentiful on earth. Had the material from which the earth was formed been hot, the arsenic would have boiled off into space.

Meanwhile others attacked the difficult problem of how the sun could have shed its angular momentum. The strange thing about the solar system is not the amount of this momentum held by the planets, for they have just enough to keep them in orbit where they are. It is the lack of it in the sun. If the solar system condensed from a dust cloud with enough spin to account for the angular momentum of the planets, the sun should be spinning 100 times faster than it is. Perhaps it did, in fact, have this momentum to begin with, but somehow shed it.

Among those who proposed a way in which this could have come about was Hannes L. Alfvén of Sweden, who said magnetic interaction between the sun and a cloud of electrically charged particles surrounding it could, over a long period of time, have produced the necessary braking effect. A somewhat similar

---

While many proposals have been made regarding the formation of the solar system, most of them are related to one of the three basic concepts shown here. In Kant's view the primordial material, drawn to itself by gravity, formed into whirling blobs. The central blob became the sun, surrounded by residual blobs that formed into planets. In the Laplace hypothesis the remaining material first formed rings, each of which then condensed into a planet. The Chamberlin–Moulton proposal was that gravitational attraction, as another star passed close to the sun, drew forth the material that then formed into planets.

explanation was proposed in 1961 by Fred Hoyle, who by this time had shelved his supernova idea. He said that, as the sun, in its formative stages, contracted within the present orbit of Mercury, it shed gas. Boulder-like chemical aggregates formed within this gas and were held fast within the wiry rigidity of magnetism generated by the sun, still very hot and rapidly spinning. As the gas and boulders spread outwards, they were still linked to the sun by this magnetism and therefore carried with them much of the sun's angular momentum.

The same principle is exploited, unknowingly, by a figure skater who is spinning fast and wishes to slow down. She spreads her arms, automatically retarding her spin, and then skates gracefully off. Likewise the spin of space vehicles is reduced, after their launching, by releasing weights attached to the vehicle by long strings. Centrifugal force pulls out the weights until the strings are taut, slowing the vehicle's spin. The weights are then cast loose and go sailing off into space, carrying the angular momentum with them.

According to the Hoyle hypothesis the sun would have to cast loose only one per cent of its mass in this way to slow to its observed spin. After flying out a certain distance the gas and magnetism would become too thin to keep a grip on the boulders, which would then be free to condense into planets.

Another version is that of A. G. W. Cameron at the Institute for Space Studies of the National Aeronautics and Space Administration. He has proposed that, for its first few thousand years, the sun was more than 100 times brighter than now and steadily ejected gas at a rate far greater than that of the 'solar wind' that today blows steadily past the earth. This outrushing gas, says Cameron, may have carried off the sun's spin.

His concept is of a very large nebula whose outer part condensed into comets. As elaborated in 1963 by Fred L. Whipple, head of the Smithsonian Astrophysical Observatory, this belt of comets must lie in a region beyond the orbit of Neptune, outermost of the 'regular' planets. Such comets are thought to consist of a mixture of ice and 'dirt'. In view of their gravitational effect on the orbit of Neptune, their combined mass, according to Whipple, is probably more than ten times that of the earth. However, they cannot be seen from the earth unless disruptions of their orbits bring them in close to the sun. Then they assume the familiar appearance of a comet with a glowing head and a tail that always extends away from the sun, regardless of the comet's true motion. The tail consists of material blown from the head by the solar wind and by the pressure of sunlight.

The belt of comets beyond Neptune seems unrelated to the vast cloud of comets that has been postulated much farther out, in the region from 100,000 to 150,000 astronomical units from the sun. An astronomical unit equals the earth's distance from the sun. Jan H. Oort of the Netherlands has estimated that there are 100 billion comets in this outer cloud, with a combined mass comparable to that of the earth. They are moving in very slow orbits around the sun and some of them may stray out as far as the midpoint between the sun and the nearest star, risking capture by the gravity of such stars. Thus Oort's picture of the solar system makes it so large that it is rubbing elbows with its neighbours, perhaps occasionally losing comets to another star, or vice versa. While Oort has proposed that these comets may be debris from the explosion of the 'missing' planet between Mars and Jupiter, the more conventional view is that they are another by-product of the process whereby the solar system was formed.

Because of the great diversity of views on the origin of our planets and sun, a conference was held in January 1962 at NASA's Institute for Space Studies, at the Interchurch Center on Riverside Drive, in New York. At the conference, attended by many leading protagonists, objections were raised to each hypothesis. Nevertheless, Cameron, in reviewing the situation, noted that most of today's theories envisaged a nebular origin, rather than a collision or near miss. The shadows that the telescopes reveal in the night sky show that the spiral arms of our galaxy are rich in clouds of gas and dust. Astronomers believe they can actually see stars forming – as, for example, during a recent seven-year period in the heart of the Orion Nebula. Likewise certain stars of the youthful, 'T-Tauri' variety can be observed driving excess dust and gas outward in a manner postulated for the evolving solar system.

The implication is that solar systems are common, but the

argument will be greatly strengthened if there is real agreement on how our own system came about. The space exploration of the next decade should enable us to narrow down the theories to a great extent. We now have samples of the moon's surface, and thus clues as to the nature of its interior. We will learn the precise compositions of other planets and their atmospheres to compare with those of our earth. However, study of our own solar system is not the only way to learn if it is unique. Another approach is to search for clues among the other stars of our galaxy. Such observations, carried out originally without reference to the question of whether or not there are planets elsewhere, led to surprising discoveries which are the subject of the next chapter.

6

The Puzzle of the 'Slow' Stars

THE quality of science that, above all else, enriches the spirits of those who pursue it is the manner in which it enables us to glimpse the orderliness of nature. One of the most thrilling experiences for the student of chemistry is his first exposure to the periodic table of elements and its revelation of the systematic structure of all matter.

A comparable disclosure in astronomy grew, in large measure, out of an ambitious, laborious and seemingly pedantic project at Harvard. In 1886 a classification of stars according to their spectral type was begun. As had already been noted, since the stars are too remote to be seen except as diffuse points of light, our knowledge of them depends heavily on their spectra. It had already been found by the Italian astronomer Father Angelo Secchi that the stars fall into certain spectral categories, somewhat as the first walker in a forest might pick out oaks, maples, birches and pines. While the stellar types shade more into one another than tree species, Father Secchi found that they clearly fit into a succession of pigeon-holes. In 1890 Harvard published a preliminary catalogue of 10,351 stars, classified according to their spectra, and by 1949 the entries in the catalogue had grown to 359,082 almost all of them processed by a single-minded lady named Annie J. Cannon. Miss Cannon, who had been made extremely deaf by a childhood illness, was equipped with an almost phenomenal memory and unusual eyesight.

Attempts to fathom the significance of the star types were made by astronomers in various lands. Those at Harvard soon guessed that they reflected temperature differences, but it remained for an Indian, Meghnad Saha, to decipher the specific messages contained in the spectra of each group. The stars classed by the letter 'O' displayed the spectral lines of highly ionized helium, silicon and nitrogen and thus were the hottest. Those stars classed as type 'M' proved to be at the cool end of the series. The types are therefore now listed in order of decreasing temperature, as follows:

O, B, A, F, G, K, M

Astronomy students keep the sequence in mind by means of the happy phrase: 'Oh, be a fine girl, kiss me!'

Meanwhile Russell at Princeton and Hertszprung in Europe had come upon another striking pattern in star classification. If the inherent brightness, or luminosity, of the stars is plotted on a graph against their types, arranged in order of temperature, some 90 per cent of all stars lie along a narrow band that has come to be called the 'main sequence'. The hotter a star is, if it lies on the main sequence, the brighter it is, the larger it is and the shorter is its lifetime. The graph that displays this relationship between luminosity and star type is known as the Hertzsprung-Russell diagram (see p. 61). When it was formulated on the eve of World War I, Russell viewed the main sequence as a highway along which stars travel in their long journey from birth to death. As will be seen in the next chapter, a different interpretation is now placed on the main sequence, but the Hertzsprung-Russell diagram has come to be, for astronomers, a testimony of order in the universe as awesome as the evidence of order on the atomic level revealed by the periodic table of elements.

After World War I a quite different type of classification was undertaken, largely by Otto Struve, heir to the most distinguished name in Russian astronomy and destined, as will be seen, to become one of the heroes of our story. Descended from a line of observatory directors, he too had studied astronomy, but during the war served as an artillery officer – a role that may have accounted for the erect bearing with which he carried himself. When the revolution came, he fought with the Whites and ultimately fled to Turkey whence he emigrated to the United States. He took his doctorate in astrophysics at the University of Chicago, working as well at the university's Yerkes Observatory in Wisconsin, of which he ultimately became director. Yet he did not cut all ties with his homeland. He served as secretary of a committee of American astronomers that sent food packages to their Russian colleagues during the great post-revolutionary famine.

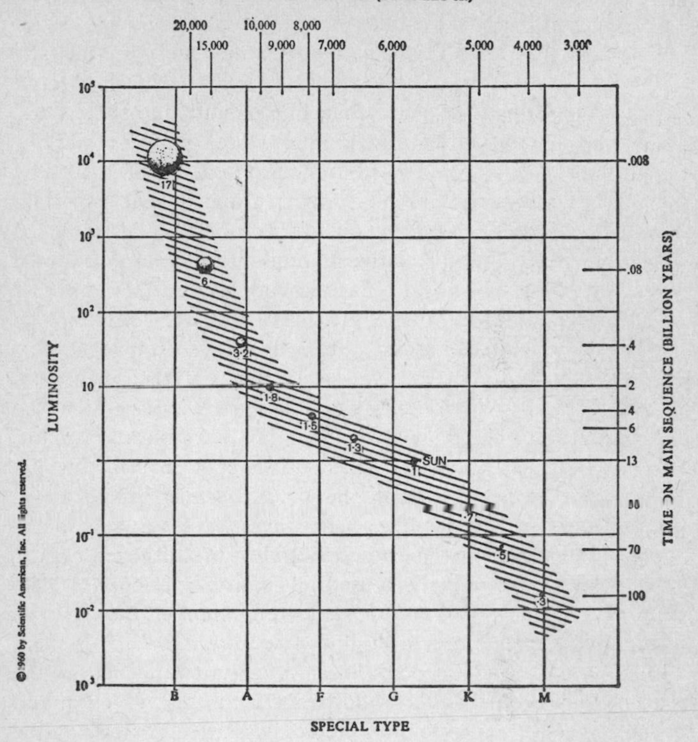

THE HERTZSPRUNG-RUSSELL DIAGRAM (SCHEMATIC DRAWING)

When stars are classified by brightness and spectral type (temperature), most lie in a band known as the 'main sequence'. Their temperature is determined primarily by their mass. The heavier the star, the hotter, brighter and shorter-lived it is. Luminosity, on the left, is shown on a logarithmic scale with the brightness of our sun as one. Temperature, across the top, is in degrees Kelvin. Star sizes are shown schematically, the figures below indicating mass, with the mass of the sun as one.

He also kept in touch with scientific developments in Soviet Russia and learned of the interest of Grigory Abramovich Shajn in the fact that stars of the same brightness and spectral type exhibit great differences in the width of the lines of their spectra. Both he and Shajn suspected that, in the light from some stars, these lines were broad because the stars were spinning rapidly.

Their suspicions were rooted in the proposal, made in 1842 by the Austrian physicist Christian Johann Doppler, that light from a moving source, such as a star (or part of a star), would be altered in wavelength by relative motion toward or away from the observer. In today's world of fast-moving vehicles this 'Doppler effect' is well known, but it was a major discovery in Doppler's time. The principle is applicable to sound waves in that they are compressed, and therefore raised in pitch, if their source is approaching. If it is receding, the waves are lengthened and the pitch lowered. This is vividly evident as a horn-blowing vehicle races past. In the case of light, the shortening of wavelength shifts lines of the spectrum towards the violet, whereas motion away, and consequent lengthening of the waves, causes a shift in the opposite direction, towards the red end of the spectrum.

In 1909 this principle was used by Frank Schlesinger of Yale University to show that certain stars are spinning. If the spin axis of a star is perpendicular to the line of sight from the viewer, then, Schlesinger reasoned, spectral lines in light from the side of the star moving towards him would be shifted in one direction and light from the opposite side would be shifted in the reverse direction. The effect would be to fatten spectral lines in combined light from the whole star.

Shajn and Struve began a statistical study of stellar rotation speeds, collaborating only by mail. They did not meet face to face until many years after their joint report had been published in the *Monthly Notices* of Britain's Royal Astronomical Society. Struve continued this work with Christian T. Elvey and it became evident that the orientation in space of the spin axes of stars was random. Thus, if there was no broadening of spectral lines, it did not mean a star was not spinning. It was possible one pole of its axis was pointing directly at the earth so that its spin produced no relative motion toward or away from the observer. A statistical analysis of spectra from a great number of stars was therefore necessary to determine if there was any relationship between spin rate and the star's other characteristics.

By 1931 Struve and Elvey had, in fact, identified such a relationship. The hotter classes of star characteristically rotate fast and thus have considerable angular momentum; the cooler stars are turning so slowly that their rotation speeds are virtually undetectable. What Struve found 'startling' was the abruptness of the cleavage between fast and slow stars. It occurs in the midst of the F-type stars. The cooler ones in that category are slow; the hotter ones are fast. Some of the O and B stars, at the hot end of the catalogue, rotate at from 250 to 500 kilometres per second at their equators, completing each revolution in a matter of hours.

Our sun, a G-type star, spins at only two kilometres per second, a point on its equator taking 25 days to make one revolution, and this seems typical of stars in this group and among its neighbouring types, the K and M stars. This was puzzling, for if these stars had condensed from clouds of dust and gas in this region of the galaxy, they should all have about the same angular momentum. The clouds themselves spin very slowly – perhaps only once for each orbit around the galaxy, which at our distance from the core takes some 200 million years. However, this spin rate increases as the star condenses.

As noted in the last chapter, a figure skater can slow her spin by spreading her arms. Conversely, she can speed it up by bringing her arms close to her sides. Her body is thus concentrated within a small radius, and for its total angular momentum to be conserved her spin rate must increase. Likewise, if a dust cloud many light years broad, that takes millions of years for a single revolution, is concentrated into a star, the latter must spin once in every few hours to retain the same angular momentum.

To Struve the example of the solar system, with its slow-rotating sun and its concentration of angular momentum in the planets, was striking. He concluded that at least part of the 'missing' angular momentum in the G, K, M and cooler F stars must reside in planets. He said there must be some 'fundamental difference' between the way in which these stars were formed and the other, fast-spinning types. The 'slow' stars of our galaxy number in the billions and Struve, in effect, proposed that all are centres of solar systems.

In 1963 Cameron at NASA's Institute for Space Studies edited an anthology of scientific papers relating to extraterrestrial life, several of which cited Struve's work as evidence for a vast number of planets. Cameron himself, however, did not give much weight to the argument. He proposed, instead (for reasons explained in the next chapter), that virtually all single stars have planets, regardless of their speeds of rotation. In his theory for the origin of the solar system, described in the previous chapter, the sun shed its angular momentum, not by transferring it to planets, but simply by discharging large amounts of gas that carried the momentum off into space.

Meanwhile, evidence that planets are common has evolved from a very different line of research. It began, in a sense, as early as 1718 when Edmund Halley, the great English astronomer, noticed that some of the stars catalogued by Ptolemy sixteen centuries earlier had shifted their relative positions. It was the first evidence that the stars are in motion with respect to one another, although they are so distant that such movement, even on a time scale of centuries, is observable only for the nearest ones.

The next great student of star motions was the German, Friedrich Wilhelm Bessel, who in 1838 found that a star in the constellation Cygnus, the Swan, known as 61 Cygni, moved back and forth among the other stars by 0·31 seconds of arc in a twelve-month cycle. He correctly concluded that the apparent motion of the star was, in fact, caused by the earth's flight around the sun. Knowing the diameter of the earth's orbit, it was a matter of simple trigonometry to calculate the distance to 61 Cygni – the first such distance measurement achieved. A few years later, in 1844, Bessel announced that, on the basis of extended observations, he had found a completely different type of cyclic motion in two of the sky's brightest stars: Sirius and Procyon. In this case the movement was that of the stars themselves, but what struck Bessel was that their paths, instead of being smooth, showed a slight wave motion or wiggle. They must each, he said, be waltzing through space with an invisible companion. As telescopes improved, both of these companions were sighted.

The effect that Bessel had identified is like that of a fat man waltzing with a skinny little girl. As they spin around one another, his motion across the dance floor seems unperturbed by her slender weight, but this obviously is not so. Close examination will show that, skinny as the girl is, her weight is sufficient to impart a slight waviness, or wobble, to his path.

The same is true of the earth-moon system. We proudly think that the moon simply circles the earth, but this is not so, for the moon is large – its radius is a quarter that of the earth – and its gravity pulls on us just as that of our planet tugs at the moon. The two bodies circle their common centre of gravity (the 'barycentre'), which lies inside the earth, but 2,880 miles from its centre. Hence, if one watched the annual flight of the earth around the sun from a distance, the path would be wavelike.

The advent of photography enabled astronomers to make far more precise determinations of relative motions among stars, since it became possible to compare pictures of one patch of sky made repeatedly over long periods. A pioneer in this work was the Danish astronomer Ejnar Hertzsprung, and in 1938 one of his co-workers, Kaj Aa. Strand, brought his technique to Sproul Observatory at Swarthmore College, near Philadelphia.

The observatory was a small one and its director, Peter van de Kamp, had decided to focus its efforts on the 54 stars known to lie within 16 light years of the sun. These are the ones close enough to display useful proper motions, although less than a dozen of them are bright enough to be seen with the naked eye. They include 28 seemingly single stars, 20 paired in binary systems, and 6 grouped into two triple systems.

Strand concentrated on the two-star system, 61 Cygni, whose distance had been measured by Bessel. Its two stars, quite widely separated, move slowly around each other, but by 1943 Strand had found that one of them circled its twin in a wiggly manner. The observation was extremely difficult, for the extent of the wiggle (0·02 seconds of arc) was far less than the width of the star's image, as it appeared on photographic plates. In part this was because such images, as a consequence of atmospheric turbulence and other factors, are greatly enlarged and blurred, covering a patch of sky from one to two seconds of arc in width. Only by using fine cross-hairs to measure distances between the centre of the star and the centres of its neighbours was it possible to detect such subtle wave motions.

In this case the wave cycle took 4·8 years. The fact that the wiggler is one of two stars, orbiting each other at a known speed and at a known distance from one another, had enabled astronomers, using Newton's laws of gravitation, to calculate how heavy they are. It was therefore possible for Strand similarly to estimate the size of the invisible object in orbit around one of them. He put it at one per cent of the sun's mass. It was a moot point whether or not such an object should be classed as a planet. It was roughly eight times the mass of Jupiter, the largest planet of which we have direct knowledge.

The work at Sproul disclosed several other bodies in our corner of the galaxy, all of them in the intermediate range, between stars and planets, where observations had previously been impossible. One of our closest neighbours, a 'red dwarf' named Ross 614, too dim to be seen except through powerful telescopes, had earlier been observed to have a wobble, and Sarah Lee Lippincott, at Sproul, was able to predict where its invisible companion would be in 1955. A photograph taken with the world's largest telescope, on Mount Palomar, confirmed the prediction. The body, with eight per cent of the sun's mass, is still large enough to crush hydrogen atoms together and, through this fusion process, shine – though dimly – in the manner of a star. The Sproul astronomers also found that another of our neighbours, Lalande 21185, has an invisible companion, this one being only slightly larger than that discovered by Strand.

Finally, the group at Sproul began studying the nearest known single star, a faint red dwarf discovered in 1916 by Edward E. Barnard and hence known as Barnard's star. It moves faster, across the back-drop of distant stars, than any other known star, traversing a path equal to the width of the moon every 180 years. It is six light years away, the only closer stars being those in the triple system of Alpha Centauri in the southern sky.

Two dozen photographs of Barnard's star and other stars near

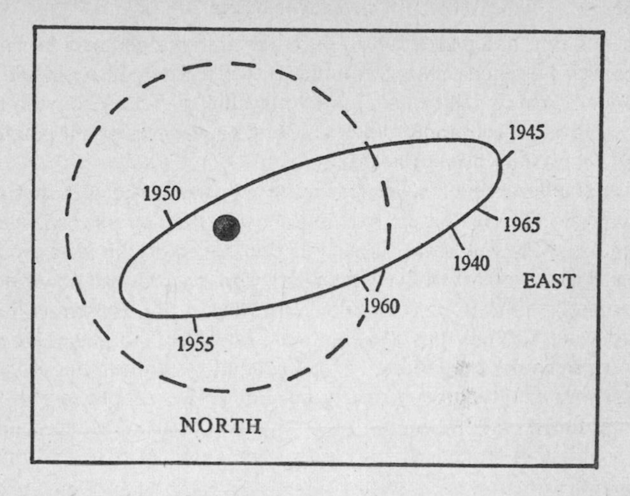

The orbit of the invisible companion of Barnard's Star, calculated by the group at Sproul Observatory, is shown as seen from the solar system, rather than from directly 'above'. Hence the plane of the orbit is foreshortened, since it is tilted to the line of sight. The planet is thought to have reached perigee, or the point nearest its parent star, in 1950. The dashed line represents the size of the star's image as it appears on photographic plates. The sun, seen from a comparable distance under similar circumstances, would appear far larger than the earth's orbit.

it had been taken from 1916 to 1919. Since 1938 the 24-inch refracting telescope at Sproul Observatory had been photographing this patch of sky routinely, providing a total of 2,413 plates taken on 609 nights. By 1956 the astronomers at Sproul were virtually certain that Barnard's star had a small companion, but they continued observing for another six years before finally making up their minds.

On 18 April 1963 van de Kamp, the observatory director, reported the discovery to the American Astronomical Society. The companion of Barnard's star, he said, takes twenty-four years to complete one orbit – intermediate between the orbital periods of Jupiter and Saturn. It imparts a wiggle to its parent star that carries it, as seen from the earth, only 0·0245 seconds of arc to either side of a smooth path. Since the star itself is very small, this means that its companion is only 1·5 times larger than Jupiter. Such an object, said van de Kamp, 'must definitely be regarded as a planet and can shine only by reflected light.'

Thus evidence has accumulated that the galaxy is richly populated with objects of all sizes between stars and planets. But where does the borderline between such fundamentally different objects lie? Van de Kamp called attention to a peculiar pair of stars discovered by Willem J. Luyten at the University of Minnesota. Instead of at least one of them being of respectable size, both have only about four per cent of the sun's mass. These midgets, known as L 726-8, glow despite their small size, but they seem to be close to the borderline. Anything smaller, in the view of van de Kamp and others, would not be sufficiently compressed, in its core, to generate the fusion reactions that cause a star to shine. An object as large as the invisible companion of 61 Cygni may glow a deep red because of its heat, but it is not a ball of incandescent gas like the sun.

There is no reason, van de Kamp wrote in the *Yale Scientific Magazine*, to doubt the existence of 'large numbers' of bodies larger than Jupiter and smaller than stars. For every star intrinsically brighter than the sun, in our neighbourhood, we see at least twenty fainter stars. If this pattern continues into even smaller sizes, they are very numerous indeed. Some, as van de Kamp pointed out, may be making their way alone through the galaxy, in which case they will be very hard to detect and, lacking any sun, would appear unlikely abodes of life, although, as set forth in the next chapter, it has been proposed that some may in fact be inhabited.

The argument that planets must be plentiful was presented by Harrison Brown, Professor of Geochemistry at the California Institute of Technology, in the 11 September 1964 issue of *Science*. On the ground that the stars we can see become increasingly numerous in smaller sizes, he estimated that, for each visible star in the vicinity of our sun, there are probably about sixty smaller bodies down to the size of Mars (which is smaller than the earth). Citing the evidence that many stars have planets, he said: 'If planetary systems are indeed extremely abundant, one might conclude with equal conviction that man is not alone – that his equivalents may occupy hundreds or even thousands of bodies within our galaxy.'

The detection of planets considerably smaller than Jupiter and therefore more comparable to the earth now seems at least possible. The wiggle imparted to a sun by such planets is very small and in a many-planet system like our own would be observable only at rare, irregular intervals when most of the planets are on the same side of the sun. Only Jupiter, in our system, is large enough to affect the sun in a manner that might be observable from nearby stars.

However, the new astrometric telescope of the United States Naval Observatory, which went into operation near Flagstaff, Arizona, in 1964, may disclose planets smaller than Jupiter simply because it can trace the paths of very small stars. The telescope, built under the supervision of Kaj Strand, who is now the observatory's Scientific Director, is termed astrometric because it is designed specifically for star-measuring. With a mirror of fused quartz 61 inches in diameter, the instrument opens up a new region of astronomic observation. Refracting telescopes, like the one at Sproul, have been the chief tools in studies of star movement but they have difficulty seeing very dim stars, in part because they must use filters to limit the colour of the light being photographed. The various wavelengths (or colours) are bent to different degrees as they pass through the lenses of such an instrument, and this effect, though slight, is enough to hamper accurate measurements. Hence a yellow filter is normally used. Reflecting telescopes, whose chief element is a large parabolic mirror, eliminate this problem, but none of the largest instruments of this type is well suited to star measuring for a variety of reasons. Thus big mirrors can detect very faint stars, but they form star images whose quality varies over the field, making it difficult to determine precise positions. This is particularly true when a telescope is short relative to its mirror size.

Another problem in using large mirrors for star measuring is that they can become sufficiently distorted, by sagging or temperature changes, to show a star out of position. The new Navy instrument has been fitted with what, for a time, was the largest quartz mirror in existence. This material was chosen because it is less distorted by day–night temperature changes on an Arizona hill top than the glass normally used for this purpose.

This instrument may at last enable astronomers to measure distances to tiny nearby stars and thus define with confidence the borderline between stars and planet-like bodies – they cannot be called 'planets' if they are not in orbit around a star. We may discover that our neighbourhood is far more crowded than we had supposed and, by observing the wiggles of these midget stars, may find that some of them have planets not much larger than the earth.

The new techniques of balloon astronomy and orbiting astronomical observatories offer a variety of opportunities for more precise star-measuring, since they minimize or obviate the blurring effects of the earth's atmosphere. The potentialities of balloon-borne telescopes have been demonstrated in recent years by such instruments as Stratoscope I, whose sun photographs were sharper than any before, and its successor, the 3-ton Stratoscope II, which has successfully scanned the spectrum of Mars. Current preparations for orbiting telescopes entirely above the atmosphere are even more exciting to astronomers, and some have suggested that, with such instruments, it may be possible to photograph planets in nearby solar systems.

This would require using some sort of disc in front of the telescope to cut off light from the star so that its planet could be seen. Such schemes have been discussed by Nancy G. Roman of the National Aeronautics and Space Administration and by Robert Danielson at Princeton. The difficulties would be formidable. Very long exposures with the most sensitive film would be needed to record anything so dim as a distant planet. The aim of the telescope would have to remain steady for a prolonged period of orbital space flight, which would demand a guidance system of extraordinary ingenuity, although the performance of Stratoscope II is encouraging in this respect.

Far more of a problem is placing the disc in front of the telescope so as to shut out the light of a star without also hiding its planets. To prevent the star's light from diffracting around the edge of the disc and masking the planets, Lyman Spitzer, Jr, head of the Princeton University Observatory, has argued that the disc must be very large, and hence extremely far in front of the telescope. For example, to detect a planet separated from its sun by one second of arc, he said the disc should be 75 metres (247 feet) wide and 10,000 kilometres in front of the telescope. A balloon could serve as the disc. However, even if the edge of the disc were designed to minimize diffraction, a telescope of about 100-inch diameter would probably be required. Both telescope and balloon would have to be orbiting the earth.

Such an arrangement might be sufficient to see a planet like Jupiter in orbit around one of the nearer stars (within 16 light years), but it would not reveal a planet like the earth, not only because the latter would be dimmer but, more important, because of its nearness to the parent star. To overcome this, Spitzer said, would require a disc five times as wide, placed 250,000 kilometres (some 156,000 miles) in front of the telescope. To position a balloon directly in front of the star, as viewed through such a distant telescope, would be extremely difficult although, he explained, this might be possible if the balloon carried a flashing light that enabled devices aboard the telescope to coach it into position and keep it there. Both balloon and telescope, in this case, would presumably be moving in a comparatively serene orbit around the sun.

Because direct observation of planets would be so difficult, astronomers have come up with various other schemes for indirect detection of a solar system. Vassily G. Fesenkov in the Soviet Union has suggested that we look for the glow produced by dust orbiting other stars, on the ground that such dust is typical of planetary systems. In our own system it produces a pyramid-shaped glow where the sun has recently set or is about to rise. Such a glow would, it is thought, be brighter than the combined lights of all planets in a solar system. Otto Struve has suggested that we might be able to detect the slight dimming of a star as a large planet moves in front of it, or that we might observe the changes in wavelengths of starlight, generated as the star is tugged back and forth slightly by orbiting planets. Such methods have one advantage: they can be applied to any visible star, no matter how far away, whereas wiggles become more and more difficult to observe with increasing distance. However,

Struve pointed out that, with present-day techniques, these alternative methods would have only a marginal chance of success.

Thus analyses of stellar motion, such as those of van de Kamp and Strand, seem to offer the best prospects for showing that planets are common. Ultimately instruments in orbit or on the moon may improve the technique. But in any case, if we find – as seems likely – that our galaxy is peppered with planets, this will not tell us how many of them are habitable. The latter question has long been the concern of a Chinese astronomer whose analysis of the problem is described in the next chapter.

7

Where to Look

OURS is a wonderful world. Of all the planets in the solar system it seems to be the only one on which life, as we know it, could survive. Mars is too dry and its air is apparently so poor in oxygen that the astronauts who land there will be unable to build fires. Venus seems far too hot. Mercury has no day–night cycle, keeping the same torrid side continuously facing the sun. The outer planets have hydrogen-dominated, frigid atmospheres in which the higher forms of earth life could not exist.

What then of planets in other solar systems, if there are any? Is the combination of factors that makes the earth habitable so remarkable that its repetition elsewhere is unlikely? A man who has probably devoted as much time to this problem as anyone is named Su-Shu Huang. He was born in Kiangsu Province, China, in 1915 and came to the United States in 1947 on a fellowship granted by the Chinese Nationalist government. He has remained there ever since. He took his doctorate in astrophysics at the University of Chicago while Struve headed the astronomy department there. When Struve moved to the University of California at Berkeley, in 1950, Huang soon followed, armed with a Guggenheim Fellowship, and he remained there for eight years before going to the National Aeronautics and Space Administration. As Struve, the Russian émigré, became more interested in the possibility of extraterrestrial life, so did Huang, the Chinese émigré. Their discussions of the subject were numerous.

In 1957 Huang sought to explain the various categories of stars: those that spin fast, those that spin slowly, those in multiple systems, and so forth. Two factors, he said, determined the nature of a star when it was born: the angular momentum for each ton of gas and dust in the cloud from which it originated, and the total amount of material in the cloud available for star-making. If there was plenty of material and moderate angular momentum, he calculated, the stuff concentrated into a large, fast-spinning star. If in this situation there was less material on hand, the star formed was smaller, it acquired planets and spun more slowly. If to begin with there was a large amount of angular momentum in the cloud, the resulting system comprised two or more stars.

Huang published several analyses of the types of stars whose planets might support intelligent life. By this time what seems to be the true nature of the 'main sequence' of star types had been recognized. It is not the highway along which stars evolve from one stage to the next. Rather it is their 'home', where they remain for a major portion of their lifetimes. An individual star does not change its position on the main sequence. Its location is predetermined by its mass, which in turn determines its luminosity, its temperature, its special type and its longevity. (See the diagram on page 61.) The main sequence is the home of a star so long as its 'fuel' is hydrogen, which is converted by a thermonuclear reaction into helium. The fact that this is by far the longest stage in a star's lifetime explains why 90 per cent of the known stars lie on the main sequence.

A star moves onto the main sequence once its internal heat and pressure, from accretion of material, become sufficient to start the hydrogen-burning. Its stay there ends when the supply of hydrogen in the core has been exhausted and, with the outflow of heat cut off, the core contracts. Hydrogen in the outer portions of the star then begins to 'burn' and increased temperature and pressure in the core make possible the reactions that convert helium to carbon. The star swells into a 'giant' or 'supergiant', becoming progressively hotter and brighter – and, presumably, destroying any life in its vicinity.

Obviously the only period in a star's history when life could evolve on one or more of its planets would be while it was on the main sequence. Only then would its heat output be stable long enough for such evolution to occur. The age of the earth is estimated at some 4·6 billion years, of which about a third was required for evolution of the chemicals that finally organized themselves into what we call 'life' (see Chapters 8 and 9). The remainder of this period then saw the evolution of life from its most primitive form to creatures able to communicate over interstellar distances.

The big stars burn their hydrogen so fast that some of them stay on the main sequence a mere million years. At the other end of the scale the small M stars remain there more than 100 billion years. However, as Huang noted, these small stars are so feeble that only a narrow zone around them offers a temperature range within which life, as we know it, could evolve. While this habitable zone around a big, hot star is much farther away from the star, it is also much deeper and more voluminous. All else being equal, the larger this zone, the greater the likelihood that there will be at least one planet whose orbit lies entirely within it.

Huang likened the situation to a bonfire in a field on a cold night. If the fire is small the zone in which people will be comfortable is narrow. If it is a big fire they must stand farther back, but the zone will be broad. The temperature extremes whereby he set the limits of his zones were not those that determine the habitat of human beings, but those necessary for the chemical reactions in any conceivable life process. Thus in the solar system Mars lies within this zone, though near its outer limit, and Venus likewise lies inside it, though near its inner limit.

While the biggest stars on the main sequence have broad habitable zones their lives are too short. While the smallest stars have very long lives their zones are extremely narrow. This leaves the moderately small stars – the smaller F types, the Gs and the larger Ks – as the most likely candidates. By a happy coincidence, Huang pointed out, those are all slow rotators and therefore, he believes, have planets. Furthermore, the sun, a G star, lies squarely in the middle of this grouping, just as the earth lies in the centre of the solar system's habitable zone.

A major problem is the fact that, at least in the neighbourhood of our sun, a large percentage of the stars are in multiple systems, most of them binary. A planet in a binary system is apt to fly in complex, frequently changing orbits, sometimes roasted on both sides at once, as it sails between the stars, sometimes lethally frigid as it reaches far out into space. However, Huang did not completely rule out the possibility of life in such systems. For

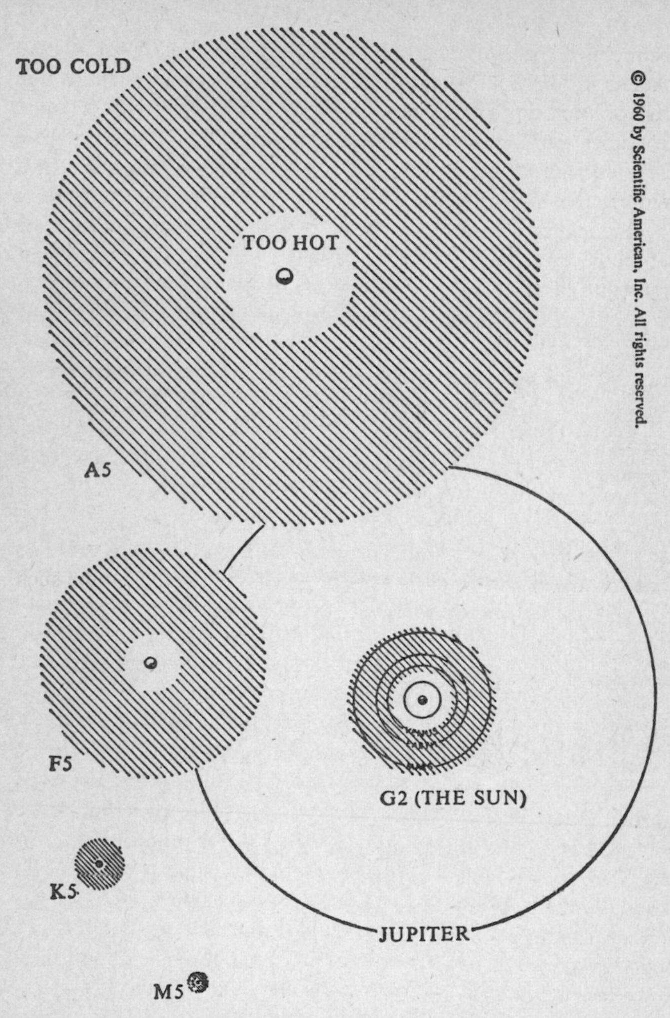

Habitable Zones (Shaded Areas)

The brightness and hence the warmth of a star determines the depth of the habitable zone around it. However, the very large hot stars such as the type A5 star at the top do not live long enough to give life a chance to develop. The sun, lower right, is a G2 star, which seems 'just right' for the emergence example, if the two stars are far enough apart, each could have planets in fairly stable orbits. The distance between them, in the case of stars similar in size to our sun, would have to be at least ten times that between the earth and sun. Or, if the stars were close enough to each other, a planet could orbit the pair in a manner that would keep it within the habitable zone. According to Huang's calculations the stars in that case would have to be separated by no more than ·05 times the earth–sun distance. That a satellite can, indeed, maintain a stable orbit despite the presence of another large body is demonstrated by the regularity of the moon's monthly flight around the earth in a region where the sun's gravity holds sway.

Binaries display an immense variety in the distances between them. Some of the ones nearest us are far enough apart to be seen as two stars ('visual binaries'). An example is the brightest of all stars, Sirius, accompanied by another star from which it is separated by some forty times the earth–sun distance. The two stars take fifty years to circle one another. Other binaries are so close that their revolutions take only a few hours. In some such cases they are detected as binaries only because the light of what appears to be a single star is periodically dimmed. This occurs when the two stars are in line, as seen from the earth, with one eclipsing the light of the other ('eclipsing binaries'). Others ('spectroscopic binaries') can be observed only because of periodic splitting of the lines in their spectra. When the two stars (which appear as one in the telescope) are moving transversely, as seen from the earth, none of their orbital motion is toward or away from us and their spectrum is normal. However, later in their orbits, when one star is approaching and the other star receding, the light from each star is shifted by the Doppler effect in opposite directions. The result is a splitting of each line in the spectrum.

These methods also make it possible to calculate how fast the

---

of life. The orbit of the earth lies close to the centre of its habitable zone. The F5 and K5 stars lie close to the upper and lower limits of star types suitable for life according to Huang's calculations. The tiny M5 star at the bottom has a very long life but its habitable zone is so small that the likelihood of there being a planet within it is slight.

stars are circling each other and hence how far apart they are. In this way Kuiper has calculated that, while the distances between binary stars are spread over a broad range, the average distance in such systems is roughly the same as that between the sun and the centre of mass of the major planets in our own system. This, again, has encouraged the belief that there is similarity between the way binaries and planetary systems are formed.

Another critical factor, in addition to the nature of the central star and of the planet's orbit, is the size of such a world. In the first place it must be large enough to retain an atmosphere at its particular distance from its sun. Heat, as every student of physics knows, causes atoms and molecules to agitate – what is known as thermal motion. The Brownian movement of microscopic particles in a fluid is evidence of this agitation. In the upper part of the earth's atmosphere, where the hydrogen and oxygen atoms of water vapour become separated, some of the hydrogen atoms, being very light, are sufficiently agitated by the heat of the sun to reach escape velocity and fly off into space. Thus the earth's gravity, which is determined by its mass, is insufficient to retain a hydrogen atmosphere indefinitely, although it is strong enough to hold the oxygen essential to our lives. Mars, with only 10 per cent of the earth's mass, has a surface gravity only 37 per cent as strong as that of the earth. Hence it has little oxygen in its air, and Mercury, with its even smaller size and proximity to the sun, has so little air that only recently has evidence of it been observed.

Some believe the earth had no air at all early in its lifetime, the lighter elements having been swept away possibly during its period of accretion. This was the conclusion, for example, of Harrison Brown at the California Institute of Technology, a leading geochemist and an enthusiast on the subject of extraterrestrial life. He was led to this belief by a puzzlc in the composition of our planet and its atmosphere, namely the rareness of argon, neon and krypton. They are barely detectable, whereas such gases as water vapour, carbon dioxide and nitrogen are comparatively common. In terms of cosmic abundances, the first-named gases should be far more plentiful. Brown's explanation is that, to begin with, all gas in the vicinity of the earth had been swept away by some process and that our atmosphere came from the chemical decay of the rocks and outgassing from volcanoes. Because water, carbon dioxide and nitrogen are components of many chemical compounds, they were included in the solid materials that accreted to form the earth, but in the world of chemistry, argon, neon and krypton are 'loners'. They shun chemical combinations and hence were scarce in the primeval rocks.

In any case the oxygen that we now breathe could not have existed in the atmosphere until its hydrogen had flown away. The reason is that oxygen atoms rush into the arms of hydrogen atoms to form water whenever free hydrogen is on hand. Therefore, for an oxygen atmosphere to appear, a planet must be small enough (as is the earth) so that over an extended period of time its hydrogen will escape into space.

It is now widely thought that most of the oxygen in our air was released by the plants of today and the past. The early stages of life presumably evolved in an atmosphere dominated by hydrogen compounds – methane and ammonia. However, Huang argued that the higher forms of life could have evolved only in an atmosphere containing oxygen because that gas makes possible a far more efficient body chemistry than hydrogen. We derive our energy primarily from the combination of glucose sugar (that is, its combination with oxygen), whereas in a hydrogen-dominated environment the glucose would have to be fermented to ethyl alcohol and carbon dioxide, a reaction that produces less than one-tenth as many calories.

In such an environment a creature must eat ten times as much to get the same amount of energy. 'Therefore,' Huang wrote in a 1961 issue of the magazine *Sky and Telescope*, 'it is doubtful that a mind such as man's would appear . . . because living creatures would be too preoccupied with seeking food.' His argument, however, rests on two unknowns: namely, whether food would, in fact, be in short supply and whether severe competition for sustenance stimulates or impedes the development of intellect.

Huang calculated that for a planet to be large enough to hold its oxygen, yet small enough to get rid of its hydrogen in time for advanced life to evolve, its radius should lie between 1,000 and 20,000 kilometres. The radius of the earth is some 6,380 kilometres. He did not argue that all planets within those limits would necessarily be habitable. In fact, we can see some that are not. Mercury, for example, is so near the sun that most of its gases are 'boiled off'. Our moon, while falling within the radius limits set by Huang, is made of material that is too lightweight. Only if it were particularly dense would a body that size have sufficient gravity to hold an atmosphere. What he sought to do was specify the outside limits within which a planet might be suitable if all other considerations were favourable.

One of these considerations is the chemical composition of the planet itself. It is hard to see how life could emerge without an abundance of the heavy elements found on our earth, particularly carbon. Hence, in 1963, Huang pointed out that the first-generation stars of our galaxy are poor candidates.

These 'Population II' stars are thought to have been formed when our galaxy was still spherical in shape. Motions within its gas clouds were random, but gradually they cancelled one another out until there was a single, residual spin that caused the galaxy to flatten into a spiral disc. Meanwhile the first-generation stars had been making the heavier elements in their cores, some of them ending their lives explosively as novae or supernovae, enriching space around them with their heavy elements. From this material the later-generation (or 'Population I') stars were formed.

From the spectra of these newer stars we can tell that they are ten times richer in metals than stars of the first generation. The latter remain distributed throughout the original sphere of the galaxy, whereas the later-generation stars, including our sun, lie in the spiral arms. Hence, when one looks upward at night, virtually all the visible stars are of later generations.

Another feature of the older stars is that few of them are in multiple systems, compared with the younger ones. This, according to Huang and his colleagues, seems to imply that such old stars have fewer planets, since double stars and solar systems seem to be products of roughly the same process. All of which points to the existence of life primarily within the arms of our galaxy – where we are – rather than beyond.

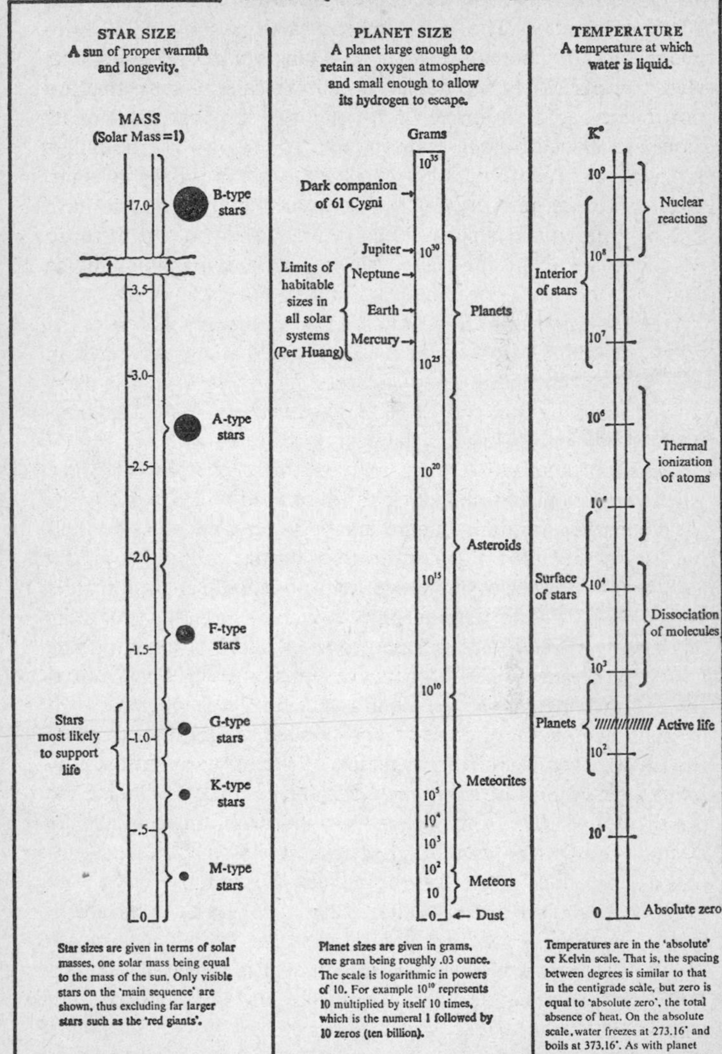

Star sizes are given in terms of solar masses, one solar mass being equal to the mass of the sun. Only visible stars on the 'main sequence' are shown, thus excluding far larger stars such as the 'red giants'.

Planet sizes are given in grams, one gram being roughly .03 ounce. The scale is logarithmic in powers of 10. For example 10^{10} represents 10 multiplied by itself 10 times, which is the numeral 1 followed by 10 zeros (ten billion).

Temperatures are in the 'absolute' or Kelvin scale. That is, the spacing between degrees is similar to that in the centigrade scale, but zero is equal to 'absolute zero', the total absence of heat. On the absolute scale, water freezes at 273.16° and boils at 373.16°. As with planet size, the scale is in multiples of 10.

Finally there is the question of uniformity in solar systems. The more uniform they are, the more likely is it that many will exhibit the remarkable combination of favourable factors enjoyed by our own planet. Without agreement on how solar systems are formed there can be no common viewpoint on this matter. There are optimists like Lloyd Motz, Associate Professor of Astronomy at Columbia University in New York, who told the 1963 meeting of the Institute of the Aerospace Sciences that planetary systems are as uniform as salt crystals. The configurations of both, he reasoned, are determined by physical laws (as suggested by Bode's 'law' expressing the systematic spacing of orbits in our own solar system). The shape of a salt crystal is the same, be it formed on earth, on Mars, or on a planet a billion billion miles distant. By the same token, he said any star ranging in size from 5 per cent smaller than the sun to 10 or 20 per cent larger is bound to have a system of planets one of which is orbiting roughly 92 million miles from the star – that is, at the earth's location. In such a situation, he added, life is virtually sure to arise.

Others, despite their interest in the possibility of extraterrestrial life, take a far less optimistic view on the uniformity of solar systems. Cameron, at NASA's Institute for Space Studies, believes there is little relationship between the classification of a star and the nature of the planets that may be orbiting that body. This is a natural sequel to his view that slow spin is not in itself an indication that planets are present. In the book *Interstellar Communication* that he has recently edited, he argues that, since there is great variety in the densities and spin rates of interstellar clouds, there will also be great variety in the distribution and sizes of planets formed from such clouds, as well as in the size of the central star. However, because these spin rates tend to be greater near the centre of the galaxy, he proposes that solar systems there will have smaller planets in more widely spaced orbits than in our part of the galaxy.

The effect of such pessimistic arguments is to reduce the number of probably inhabited planets in our galaxy from the billions to the millions, most of them very distant. Huang sought to analyse which of our neighbours, within 16 light years, might provide light and warmth to some race of intelligent beings. Our nearest neighbour, Alpha Centauri, which is 4·3 light years away, is a triple system within which habitable orbits seem unlikely. Most of our other neighbours are dwarf-like M stars that produce little heat. Huang did not disqualify them entirely. If, by good fortune, there should happen to be a suitable planet orbiting within the narrow habitable zone around such a star, it would have an extremely long period of stability in which life could emerge. 'Consequently,' he wrote, 'living organisms there can evolve to a very high form.' This might come about, in particular, since the M dwarfs are so very numerous, but he thought this unlikely among the stars of this type in our immediate vicinity.

When these, and members of multiple systems and fast-moving stars that are transients in our part of the galaxy, are all eliminated, he found that only two stars are left, apart from our sun. Both are some 11 light years away and are dwarfs with an intrinsic brightness about one-third that of our sun. One, Epsilon Eridani, is a K-type star. The other, Tau Ceti, is of the G class, as is the sun.

In acknowledging the role of Otto Struve in stimulating this analysis, Huang said that Struve had several times pointed to Tau Ceti as the possible abode of a life-bearing planet because of its resemblance to the sun. The two astronomers were not the only ones to whom it occurred that life might exist near these two stars, as will be seen later in the chapter on Project Ozma.

A radical departure from Huang's analysis is the proposal of Harlow Shapley that life may exist, without benefit of a sun, on some bodies intermediate between stars and planets. As noted in the last chapter, the possibility that such objects exist individually, like the single stars, has been greatly increased by the discovery that some of them are in orbit around our stellar neighbours.

Shapley elaborated his argument in a paper published in 1962 by the National University of Tucumán in Argentina. He pointed out that our theories of planet and star formation allow for the accretion of all sizes of body below a certain ceiling. No star can apparently survive that is more than about seventy times as heavy as our sun because the expansive pressure of radiation from within it would blow the star apart. As noted earlier, big stars are scarce, medium-sized stars like the sun are more numerous, and tiny stars are very plentiful. Hence the bodies too small to shine as stars must be even more numerous. 'To me,' he wrote, 'they seem inevitable.'

Where are these 'darkies of interstellar space'? Everywhere, Shapley said. They may be more numerous in the spiral arms of the galaxy or in the great nebulosities than in the 'stellar deserts', he added. There is apparently none near enough the solar system for its gravity to have an observable effect on the outer planets. 'Myriads' of these bodies, Shapley continued, 'are not orbitally obedient to any star – these in addition to the planets of all sizes that are immediately subservient to stars.'

But how could that 'chemical delicacy' which we call life arise and survive on such bodies? Shapley pointed out that Jupiter, being five times farther from the sun than the earth, gets little heat from our star, but it is so massive that the heat output of its interior must be considerable. The heat within the earth is generated chiefly by the radioactive decay of materials in its interior, but gravitational contraction also is a contributor. A planet many times the size of Jupiter would generate so much heat that its surface rocks would be molten, but, said Shapley, there must be a critical size whose surface temperature is such that its crust will be solid and water will exist in liquid form. The size must be at least ten times that of Jupiter, he added.

Such a body would not enjoy the solar radiation that seems to have played an important role in stimulating the chemical evolution that led to the emergence of life. However, the lightning discharges that are also thought to have figured in this process would be present, as would an abundant outflow of energy. This internally produced energy is manifest on earth chiefly in volcanoes, hot springs and earthquakes. On so large a body it would be far more evident. The surface would glow brightly in the deep infrared, and this might be exploited by organisms, both for vision and for photosynthesis – the manufacture of carbohydrates by plants. Photosynthesis as we know it would be impossible in the absence of sunlight.

Thus, scattered through the universe, said Shapley, there must be countless bodies of this sort on which the conditions are suitable.

Then something momentous can and undoubtedly does occur [he said]. The imagination boggles at the possibilities of self-heating planets that do not depend, as we so, on the inefficient process of getting our warmth through radiation from a hot source, the sun, millions of miles away. What a strange biology might develop in the absence of the violet-to-red radiation!

One difficulty, as Shapley recognized, is the enormous gravitational force on the surface of such a body. Presumably few if any earth creatures could endure it. Perhaps only oceanic life could evolve.

Ultimately, he suggested, we may find ways to detect such objects in our corner of the galaxy. Photographic emulsions very sensitive in the deep infra-red might provide one method, or the use of electronic image-intensifiers of extraordinary sensitivity, such as are already being developed for astronomical work. Also, 'in a century or so', he said, star measuring methods may have advanced to the point where it will be possible to detect the long-range gravitational effect of such bodies on neighbouring stars. However, the use of large radio telescopes seems the best hope, in Shapley's view, since such bodies should emit considerable radiation in the radio part of the spectrum (in wavelengths measured in centimetres and millimetres).

Thus, Shapley claimed, the nearest life beyond the solar system may not be on a planet orbiting a star, but on one of these lonely bodies. Ultimately its presence may be revealed through its production of heat-generated radio waves or, he said (hinting at the possibility of intelligent life), be 'otherwise made manifest'.

8

The Origin of Life: Creation or Evolution?

'WHERE did we come from?'

Men have asked themselves this question since primitive times but, as in so many other fields, science did not come forth with a plausible answer until the mid-twentieth century. New discoveries in biochemistry have suggested the steps whereby life could have arisen from inanimate matter. They also imply that the same thing has occurred elsewhere.

Thus science, having relegated to absurdity the idea that we are in any way central to the universe, and having made it appear unlikely that our planet is unique, now proposes that our most proud quality, life itself, is not unique to the earth.

Not that this idea is new. In ancient times Anaximander of Miletus was prophetic both in his argument that the number of worlds is infinite and in his view on evolution. He suggested that life originated in sea ooze and adapted itself to a multitude of environments. Even man himself, he said, was descended from sea creatures. However, such ideas conflicted with the view that so marvellous and mysterious a phenomenon as life could only come about through some form of divine action. Even at the time of Anaximander, some 600 years before Christ, this concept of the miraculous simultaneous creation of all species was deeply rooted in many (though not all) of the world's religions.

Indeed, it persists today in the minds of some fundamentalists and led in 1925 to the famous trial of John T. Scopes, a biology teacher, for violating a Tennessee law that forbade the teaching in public schools of any concept of man's origin that contradicted the Bible.

Yet observing men, from earliest times until a century ago, believed that life could arise from inanimate matter. They noted that maggots and worms appear, as though from nowhere, in rotting material and regarded this as evidence that life can be formed spontaneously. This idea of 'spontaneous generation' was accepted by such pillars of seventeenth- and eighteenth-century science as Newton and Harvey, the discoverer of the circulation of the blood.

By the seventeenth century, nevertheless, at least one man attacked the problem of spontaneous generation scientifically. He was Francesco Redi, court physician to Ferdinand de' Medici of the famous Florentine family. Redi showed that the white worms in rotting meat hatch from eggs laid there by flies. Using 'controls', like a modern researcher, he placed identical pieces of meat in two jars. One jar he covered with gauze; the other he left open. The flies, plentiful in seventeenth-century Florence, swarmed about the jars, but could not land on the covered meat. The worms, of course, appeared only in meat accessible to the flies.

Yet Redi's experiment did not destroy the idea of spontaneous generation. The seventeenth-century Belgian, Jean Baptiste van Helmont, who did some of the earliest experiments in physiology, believed that the key to life lay in fermentation and proposed methods for the generation of scorpions and other creatures that today seem preposterous. His most famous was his recipe for mice:

> If one stuffs a dirty shirt into the orifice of a vessel containing some grains of wheat [he wrote], the fermentation exuded by the dirty shirt, modified by the odour of the grain, after approximately twenty-one days brings about the transformation of the wheat into mice.

During Redi's lifetime, the debate took a new turn when a draper's apprentice in the Lowlands, Anthony van Leeuwenhoek of Amsterdam, turned the recently discovered microscope to biological use and gazed upon the wonderful world of bacteria and protozoa. To many, such organisms seemed to arise spontaneously, although Leeuwenhoek himself suspected that they drifted through the air. The latter view was strengthened in 1765 when Lazzaro Spallanzani, an Italian naturalist, put soup, of a kind in which bacteria thrive, in a flask whose neck was drawn out to a fine tip so that it could be quickly sealed by melting. While the soup was boiling, Spallanzani sealed the flask and, no matter how long it was kept, no clouding appeared to indicate the activity of bacteria.

When François Appert, a French businessman, heard of this experiment, he realized its significance in food preservation, and the canning industry was born. Still the idea of spontaneous generation refused to die. In England a Jesuit priest, John Tuberville Needham, rejected the significance of Spallanzani's experiment. In fact he argued from the Bible that all life, in a sense, arose spontaneously. He pointed out that, according to Genesis, God did not create life directly but commanded the earth and sea to bring it forth:

> And God said, Let the earth bring forth grass, the herb yielding seed, and the fruit tree yielding fruit after his kind . . .
>
> And God said, Let the waters bring forth abundantly the moving creature that hath life, and the fowl that may fly above the earth in the open firmament of the heaven . . .
>
> And God said, Let the earth bring forth the living creature after his kind, cattle and creeping thing, and beast of the earth after his kind: and it was so.

Only man, in this account, was directly created by God. Father Needham argued that the earth and sea are still carrying out God's command.

Early in the nineteenth century experiments with electricity allegedly created organisms in the laboratory. Andrew Crosse reported that, when he soaked porous stone in a mixture of hydrochloric acid and potassium silicate, then ran an electric current through the stone, fearsome-looking creatures of microscopic size emerged. It is now assumed that the creatures, though unobserved, were there to begin with.

Meanwhile Buffon, the French naturalist who invented the collision theory to explain the origin of the solar system, came forth with an ingenious explanation for the seemingly spontaneous appearance of life. Living matter, he said, consists of 'organic molecules' that, during the process of decay, can rearrange themselves to form new organisms from recently deceased material.

Buffon's countryman, Laplace, in addition to suggesting a dust-cloud origin for the solar system, proposed that the plants and animals of the earth were brought into existence by the beneficent action of sunlight. He reasoned that the same thing had occurred

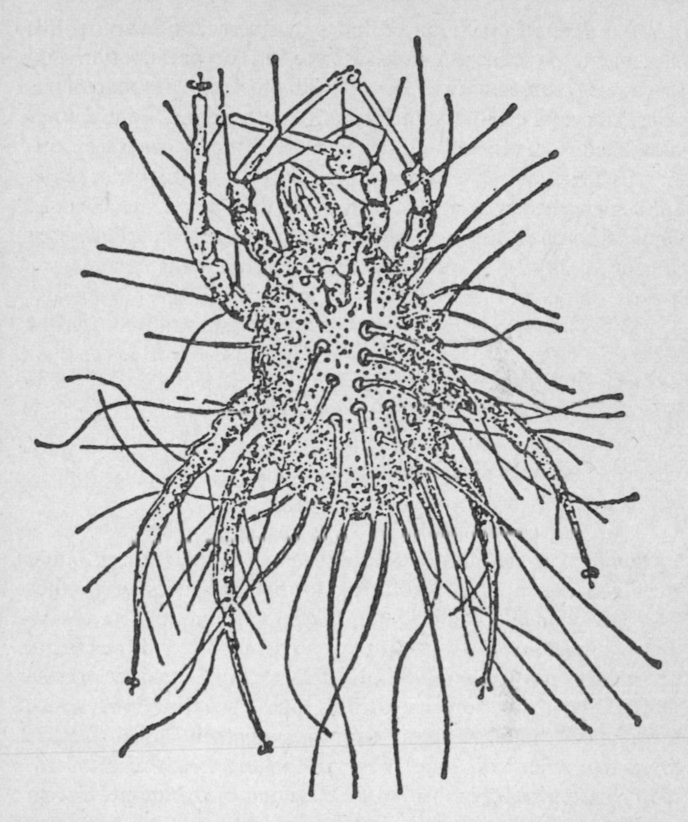

This microscopic creature was alleged by Andrew Crosse to have been synthesized by the action of electricity on a porous stone soaked with hydrochloric acid and potassium silicate. The organism, named Arcarus electricus, was shown in H. M. Noad's *Lectures on Electricity* (London, 1849).

on other planets despite the different temperatures existing there. The life that would arise on such planets, he said, would be suitable for that environment.

However, the question as to whether or not lowly forms of life can arise spontaneously did not provoke passionate debate until, in the theory of evolution, it was suggested that all species, including our own, are descended from some primitive creature. The argument then brought into conflict the ideas of two of the scientific giants of the nineteenth century: Louis Pasteur and Charles R. Darwin.

Darwin's theory of evolution was rooted in the classification of species that had been going on since the days of the great Swedish botanist Linnaeus, a century earlier. Naturalists, travelling to distant lands and oceans, had vastly enlarged the catalogue of known plants and animals. More significant, when the internal, as well as external, characteristics of these organisms were studied, it became clear that there was some form of 'system' in the plant and animal kingdoms. The situation was somewhat analogous to that confronting those who classified the stars, the basic elements, or, in our time, the subatomic particles. In each case there was clearly some inner meaning to the orderliness of nature, the problem being to fathom what it was.

Some of Darwin's predecessors, in particular the chevalier de Lamarck, had guessed that the grouping of species into family trees was the result of evolution from one species to another. Lamarck thought evolution had taken place though the adaptation of plants and animals to their environment and the passing on, to the next generation, of these acquired characteristics. Darwin then came forward with the idea that new types emerge through occasional, random variations, or errors, in heredity and are then weeded out through survival of the fittest.

In looking back down the long avenue of evolution, Darwin concluded that, in the dim past, there must have been some single, primitivc form of life from which everything else arose. He then logically asked: Whence came that original species? In a recently discovered letter, thought to have been the last that he dictated and signed before his death, Darwin said the knowledgc of that time (1882) was so meagre that any serious attempt to explain life's origin would be premature. He conceded that he knew of no reliable experiment demonstrating the spontaneous generation of life, but he reaffirmed his belief that life itself must have evolved, as did the species that came later. The 'principle of continuity,' he wrote, 'renders it probable that the principle of life will hereafter be shown to be a part, or a consequence, of some general law . . .'

Darwin recognized the difficulty confronting any theory whereby life came about through the coming together of the key chemicals of the life process. These chemicals cannot exist in the world of today except inside of living things. In the open, so to speak, they are quickly gobbled up, either in chemical reactions or by living organisms. However, Darwin pointed out that conditions in the past may have been different. In a letter prior to 1871 he wrote:

> It is often said that all the conditions for the first production of a living organism are now present, which could ever have been present. But if (and oh! what a big if!) we could conceive in some warm little pond, with all sorts of ammonia and phosphoric acid salts, light, heat, electricity, etc., present, that a proteine compound was chemically formed ready to undergo still more complex changes, at the present day such matter would be instantly devoured or absorbed, which would not have been the case before living creatures were formed.

Darwin was gifted with remarkable scientific intuition, and the extent to which he anticipated modern thinking on this subject is particularly striking, as will be seen in the next chapter.

Whereas Darwin's genius was in his flashes of insight and imagination, Pasteur was passionately devoted to the experimental method. While the French scientist was still a young man, Rudolph Virchow in Germany established the doctrine that the cells constituting all life, from one-celled organisms to man, cannot arise except as the offspring of other cells. The stream of life, both in terms of an individual's growth and the flow from generation to generation, was thus a succession of cell divisions. This seemed incompatible with spontaneous generation, but did not change the beliefs of many highly respected scientists.

Among these was Félix A. Pouchet, Director of the Museum of Natural History in Rouen, France. He believed that some constituent of the air, such as oxygen, was the critical factor in stimulating the spontaneous appearance of bacteria in material ripe for decay.

Pasteur said that such spontaneous generation was impossible. Life, even as lowly as bacteria, could only arise from other life. The organisms responsible for fermentation, he argued, were suspended in the air. In 1860, to prove his point, he set forth from Paris with seventy-three sealed flasks, each containing a broth that was fermentable but had been sterilized by heat. His first stop was at Arbois, near his father's tannery in the vicinity of the Jura Mountains, where he opened twenty of the flasks, allowing fresh air to rush in, then sealed them again. Eight of these containers later showed signs of fermentation.

Pasteur and his helpers then climbed to the summit of Mont Poupet in the Jura, some 2,800 feet above sea level, where another twenty flasks were opened in the same manner. Of these, five produced evidence of organic activity. He then went to Chamonix, at the foot of Mont Blanc, highest peak of the Alps, where he hired a guide and mule to carry his equipment up to the Mer de Glace, the great glacier on the side of that mountain. To make sure no bacteria were swept into the flasks except those in the air itself, he broke the sealed neck of each flask with sterile nippers, holding both high over his head, then sealed the neck again by melting the glass with a flame. Of the twenty flasks opened on the Mer de Glace, only one displayed signs of bacterial contamination.

The results showed, Pasteur said, that the decay was produced by organisms that were scarcer at high elevations than at low levels. In no case would one expect all the samples to be infected, any more than one would expect all those encountering someone with a cold to catch it.

Meanwhilc Pouchet was doing some collecting of his own, exposing samples to the air on the plains of Sicily, on Mount Etna and at sea. His results were quite different from those of Pasteur. He said they showed that all air is 'equally favourable to organic genesis', whether it be in the midst of a crowded city, out at sea or on a mountain top. When Pasteur's results became known, Pouchet decided to outdo him at his own game. He and his associates organized an expedition into the Maladetta Mountains, highest of the Pyrenees, and opened a series of flasks on the edge of a glacier higher than Pasteur's loftiest experimentation site.

The solution in each flask was a decoction of hay that was ripe for decay, and the contents of all the flasks fermented.

Pasteur charged that Pouchet had contaminated each flask by breaking its neck with non-sterile nippers. At first, sentiment in the Paris Academy of Sciences was on Pouchet's side and a commission was formed to resolve the dispute, which by now had shaken much of French society. The commission wished to settle the matter through a single, joint experiment, but Pouchet refused. Meanwhile, however, the tide had turned and the scientific community had begun to accept Pasteur's results as irrefutable. The climax came on the evening of 7 April 1864 when Pasteur was invited to give one of the newly inaugurated evening lectures on science at the Sorbonne.

When Pasteur, with his stubbly beard, stepped to the rostrum he saw before him the leading lights of Paris – novelists like George Sand and the elder Dumas, scions of Parisian society, and a multitude of scientists. In a deep, firm voice he noted that they were living in a time of great debates; was man created several thousand years, or several thousand centuries, ago? Are the species of plants and animals fixed in form, or do they evolve from one type to another? (Darwin's *Origin of Species* had been published only five years earlier.) Pasteur said he would direct his remarks to one such question, 'Can matter organize itself on its own? In other words, can beings come into the world without parents, without ancestors?'

The controversy on this subject, which had spilled over into the public domain, he said, was essentially between 'two great currents of thought as old as the world, and which, in our time, are known as materialism and spiritualism.'

'What a victory, gentlemen, for materialism,' he continued, if it could be shown that matter can organize itself and come to life. 'Ah! if we could give [to matter] that other force which is called life . . . what need to resort to the idea of a primordial creation, before whose mystery one must indeed bow down? What need for the idea of a God-creator?'

Pasteur, a devout Catholic, then ridiculed a book published a few years earlier by Jules Michelet, a rather fanciful writer, who pictured life originating in a nitrogen-rich drop of seawater – a bit of mucus or 'fecund jelly' that, in perhaps 10,000 years, evolved 'to the dignity of insects' and in 100,000 years to the monkeys and man. What most irked Pasteur was Michelet's statement that this concept had taken on new strength, with 'much éclat', as a consequence of Pouchet's experiments.

Pasteur showed the audience a flask, part of which had been drawn out to a point. Such a flask, he said, containing fermentable material, had been opened at the tip of its point four years ago, but because it was shaped so that no dust could enter, there had been no fermentation of the material inside. Spontaneous generation, whether by contact with pure air or by any other means, was out of the question, he said:

> No, there is no circumstance known today whereby one can affirm that microscopic beings have come into the world without germs, without parents resembling themselves. Those who claim it are the playthings of illusions, of badly done experiments, tainted with errors that they did not know how to recognize or that they did not know how to avoid.

From the perspective of a century René J. Dubos, French-born bacteriologist at the Rockefeller Institute in New York, who has studied the work of both Pasteur and Pouchet, has come to the conclusion that both, in a sense, were right and both were wrong. He believes it was, in fact, pure air that activated the bacteria in Pouchet's flasks, not by generating them spontaneously, but by bringing to life bacterial spores already in the flasks. What neither Pasteur nor Pouchet realized, he says, was that bacteria can often withstand high temperatures and did not die when Pouchet originally heated the flasks.

A decade after his lecture at the Sorbonne, Pasteur dealt another blow to the idea that life can spring from non-living matter. He said there is a peculiar quality to the chemical substances in living things that sets them fundamentally apart from non-living substances. He referred to the work that first brought him to the attention of the scientific world – his study of the manner in which tartaric acids (derived from fermented grapes) twist light waves (that is, change the plane of polarization either to the right or left). The effect is produced by dissymmetry in the structure of molecules that are related to life, he said. It is not a matter of chemistry, he added, but rather the manifestation of a 'force' that has its roots in the dissymmetry of the universe itself.

So effectively did Pasteur discredit the adherents of spontaneous generation that only a few brave souls dared speculate about how life might have originated. The rest clung to the view that some miraculous 'spark' was needed to 'breathe life' into the first living thing (or followed the Oriental philosophy that life has always existed).

Among the few who dared say otherwise was Alexander Winchell, first Professor of Geology at the University of Michigan. When he moved to a similar post in Tennessee, according to an account by Melvin Calvin, his teachings on the origin of life led to such virulent attacks that he was asked to resign. He replied that he would do so only if the trustees announced the reason, which they declined to do. Instead his job was abolished and he returned to Michigan, where his *Sketches of Creation* appeared in 1870.

Early in this century there were scattered attempts to discuss how life might have originated on the primitive earth, but not enough was known of the chemistry of life (biochemistry) or of the probable history of the earth to make possible the development of a persuasive theory. In Sweden Svante August Arrhenius, one of the first winners of the Nobel Prize in chemistry, tried to get around the problem by proposing that the earth's life came from elsewhere in the form of spores wafted across interplanetary or interstellar distances by light pressure. When the hypothesis was put forth at the turn of the century, Clerk Maxwell had already discovered that electromagnetic radiation (that is, light) exerts a pressure, although it is so weak as to be unobservable under normal laboratory conditions.

The hypothesis of Arrhenius became known as the theory of panspermia – the idea that spores of microscopic life are carried upward from one planet by air currents or volcanic eruption and are then lifted by electrical effects to a sufficient height to become free of the atmosphere and subject to light pressure, such as that of sunlight. Thus the seeds of life would be carried to all parts of the universe.

The trouble with the panspermia idea was that even bacterial spores that survived boiling in Pouchet's flasks would probably be killed soon after they left the protecting envelope of air around the earth by ultra-violet rays from the sun (particularly rays of wavelength shorter than 3,000 angstroms). As calculated by Carl Sagan, then at the University of California in Berkeley, such spores could not even survive the journey from the earth to Mars, our nearest neighbour outward from the sun. However, Sagan believes the ultra-violet rays are so much weaker at the distance from the sun of planets like Uranus and Neptune that the panspermia hypothesis is not ruled out so far as they are concerned, even though it is not applicable to the origin of our own life.

It is probably more than a coincidence that contemporary discussion of the manner in which life could have come into being on our own planet was initiated by two students of the writing of Friedrich Engels, co-founder of communism, who a half century earlier had approached the subject from a materialistic point of view. One of them was Aleksandr Ivanovich Oparin, whose classic work, *The Origin of Life*, appeared in 1936 in the Soviet Union and was translated into English two years later. The other, J. B. S. Haldane, began his career in a manner that would have destined most men to become members of the British 'Establishment'. He went to Eton and Oxford and served with distinction in the Black Watch regiment during the First World War. His interest in biochemistry was, in a sense, thrust upon him, for he was gassed during the fighting and then became a specialist in the effects of poison gas. Soon after becoming a reader (instructor) in biochemistry at Cambridge University, he figured in a controversy strikingly reminiscent of a fictional one placed at that seat of learning in C. P. Snow's novel *The Affair*. At the time, Snow was a student at the University of Leicester, but the incident was still reverberating in the halls of Cambridge when Snow came there three years later.

Because he had been named co-respondent in a divorce case, Haldane, in 1925, was asked to resign from the university. He refused to do so on the grounds that his actions had not been immoral and that his personal affairs had nothing to do with his qualifications as a teacher. The case was brought before the Sex

Viri (or Six Men), a university tribunal which consisted of the masters of three colleges and three professors, with the Vice Chancellor as chairman. They reaffirmed his discharge, but again Haldane appealed, this time to a court that, it was said, had not been convened in a century. Its members included a distinguished jurist, the Member of Parliament for the University, the Provost of Eton and two well-known scientists, all of them Cambridge graduates. Meanwhile, men like Bertrand Russell and G. K. Chesterton, a leading Catholic layman as well as a novelist, spoke out for Haldane on the grounds of academic freedom. The higher tribunal promptly reinstated him.

For the rest of his life, Haldane was a thorn in the hide of the Establishment – a witty expositor of science (some called him the George Bernard Shaw of science journalism). A brilliant scientist in his own right, he was, for a time, Chairman of the Editorial Board of the *Daily Worker* in London. He later ceased his activities as a Communist, apparently because of the suppression, under Stalin, of free scientific inquiry, particularly in his own field of genetics.

It was early in this rather unconventional career – in the 1920s – when Haldane suggested that the primitive atmosphere of the earth contained ammonia, carbon dioxide and water vapour, but little or no oxygen. He reasoned that all of the carbon now tied up in coal deposits and other remains of former life had once been a part of the atmosphere as carbon dioxide. The amount of this carbon is such, he said, that all the oxygen now in the air must have been combined with it in this primeval carbon dioxide. In other words, there was no free oxygen before the plants began chewing up the carbon dioxide to make carbohydrates. Only in the absence of free oxygen could chemical building blocks have survived long enough to hook up into the complex molecules that make life possible. Otherwise they would have reacted with the oxygen – that is, they would have oxydized.

The primary building blocks, Haldane proposed, were synthesized by the action of solar ultra-violet light shining on a mixture of carbon dioxide (as a source of carbon atoms), ammonia (a source of nitrogen atoms) and water. The organic compounds differ from other chemicals (inorganic compounds) in that they are built around carbon atoms, usually linked in a chain. They are called organic because it was originally thought that they were peculiar to material that was alive – or had been alive, as in coal or petroleum. When in 1828 Friedrich Wöhler synthesized urea from purely inorganic materials, it was shown that organic substances can be produced independently from the life process.

It was just a century after Wöhler's synthesis that experiments, carried out in Liverpool, England, seemed to support Haldane's idea that such substances were first formed by the action of ultra-violet on carbon dioxide, ammonia and water. The experiments, he wrote in 1928, showed that 'a vast variety of organic substances are made, including sugars and apparently some of the materials from which proteins are built up.' Today, he pointed out, 'such substances, if left about, decay – that is to say, they are destroyed by micro-organisms. But before the origin of life they must have accumulated till the primitive oceans reached the consistency of hot dilute soup.'

In the modern world only a small fraction of the sun's ultra-violet light reaches the surface of the earth, the rest of it being absorbed by the atmosphere, particularly by oxygen in its three-atom form (ozone). But, as Haldane pointed out, if the early earth had no oxygen in its air, there would have been plenty of ultra-violet to synthesize organic compounds. As evidence that the first forms of life evolved in an oxygen-free environment, he cited the primitive, and apparently vestigial, life forms of today that cannot survive in the presence of oxygen, such as the bacteria causing tetanus and gas gangrene. Furthermore, he said, some of the higher species, which tend to recapitulate their evolutionary histories during embryonic development, also get along without oxygen for the initial days after conception. Embryo chicks at first require little or no oxygen, he claimed, deriving their energy by fermenting sugar into lactic acid like the bacteria that turn milk sour. Some mammals do likewise, 'and in all probability you and I lived mainly by fermentation during the first week of our prenatal life.'

Likewise, he said, there is little variation in the fermentation processes used by various species, whereas the oxygen-consuming processes from which higher forms derive their energy are most diverse, suggesting that they represent a later stage of development.

Haldane's concept, like that of Lucretius, was of synthesis through random interactions over a very long period of time. With nothing to consume them, the organic compounds of his soup continued to become thicker and more varied until huge, complex molecules appeared that could produce copies of themselves if they were immersed in a specific mixture of smaller molecules upon which they could draw. He likened the situation to that of a virus which can reproduce only if immersed in a particular cocktail of molecules; namely, the type of cell that that virus infects.

The next step was to enclose the self-replicating molecules in a container holding the appropriate substances – that is, a cell. Haldane observed that some oily films, as on the surface of water, tend to roll into spheres enclosing a small amount of the water. Some of these enclosed droplets would, sooner or later, have contained the appropriate cocktail. 'When the whole sea was a vast chemical laboratory,' he wrote, 'the conditions for the formation of such films must have been relatively favourable.'

Oparin pointed to another phenomenon as a possible starting point for cell formation. When a variety of proteins and other large molecules are mixed in solution, they sometimes form into droplets with peculiar properties. Some of the molecules migrate to the surface, forming a skin, and there is also segregation of material deeper within the droplets. Oparin reported that such 'coacervates' are able to absorb certain organic substances from the surrounding fluid, while rejecting others – an important property of living cells. The droplets swell with newly ingested material until they reach a certain size, then divide, much like a drop of water. Thus, in Oparin's scheme, reproduction appeared before the system for passing on the blueprint of structure and function – what we call genetics. However, today many believe the wonderful chemistry of genetics had to evolve before reproduction was possible.

Nor has everyone agreed that the open sea was a likely place for the critical period of life's evolution. Some, such as J. D. Bernal at the University of London and Shiro Akabori at Osaka

University in Japan, have argued that the constituents must have clung to clay or mineral surfaces to remain in contact long enough for chemical reaction. Akabori's laboratory has, in fact, carried out experiments to demonstrate how the white clay used in making porcelain could have served this purpose. Shallow ponds and river estuaries were proposed as ideal sites.

More attractive to many was the picture painted by Haldane: a primeval ocean, lifeless, but murky with organic chemicals of incredible variety, heaving under a lifeless sky and beating upon lifeless shores. Then, at last, an oil-skinned droplet is formed with all of the necessary ingredients, including the ability to duplicate itself. With no competitors, this organism spreads like wildfire until checked by limitations in the food supply. Further progress is then limited until one of the organisms stumbles on the way to produce its own food, by means of chlorophyll, using solar energy for photosynthesis.

This would free life from dependence on the random synthesis of carbohydrates in the ocean and set in motion the biological evolution that has produced the marvellous diversity of today – plant and animal species that fill every nook and cranny of the environment, on land, in the sea and in the air, where life can be sustained through chemical ingenuity, or dexterity or brute force – or finally through intelligence.

Oparin's concept of the origin of life was much like that of Haldane, but it differed in one important respect: he proposed that the primitive atmosphere was a mixture of methane (the principal component of natural gas used in our homes), ammonia, water vapour and hydrogen. He also saw that a prolonged period of chemical evolution must have preceded the appearance of what unequivocally could be defined as 'alive'.

> It must be understood [he wrote] that no matter how minute an organism may be or how elementary it may appear at first glance it is nevertheless infinitely more complex than any simple solution of organic substances. It possesses a definite dynamically stable structural organization which is founded upon a harmonious combination of strictly co-ordinated chemical reactions. It would be senseless to expect that such an organization would originate accidentally in a more or less brief span of time from simple solutions or infusions.

Thus, when we gaze through a microscope at the tiny rods, spheres and corkscrews that we call bacteria, they seem the picture of simplicity, yet their internal chemistry is almost as complex as the activity of a throng of human beings, all with their various tasks to fulfil so that their society, as a unit, may thrive. In a typical cell there are some 10 billion molecules and, since this is roughly the total number of cells in our bodies, it has been argued that the evolution of the first cell from the primordial soup must have taken about as long as our evolution from one-celled organisms.

Not only did the primordial chemicals have to build themselves up into the huge molecules (comprising thousands of atoms) that are essential to the life processes; those molecules had to discover the efficient ways of working together that consitute life itself. Returning to the analogy of molecules in the cell as millions or billions of people, they had to become organized into a great cooperative enterprise, instead of competing haphazardly against one another. Furthermore, the mechanism of heredity had to evolve before there was any genetic system to perpetuate the organization once it had been achieved.

That all of this took place as a consequence of random interactions between atoms and molecules seems, at first glance, incredible. It is like the oft-cited illustration of the monkey pecking randomly at the typewriter. Since he must eventually type all possible arrangements of letters and spaces, given enough time he will ultimately 'write' *Hamlet*. The only problem is to give him enough time (in billions of years). But, as will be seen in the next chapter, the construction of the big life-molecules is not entirely random. Because of certain characteristics of the atoms involved, and consequent laws of chemistry, the evolution of such substances is bound – or at least apt – to follow paths now being explored by the biochemists.

The other factor that, if it can be comprehended, makes such evolution appear credible is the time span available. When Oparin first wrote his book, he did not have at his disposal either the information that we now possess as to the timetable of the earth's history or our modern knowledge of life's intricate chemistry. By determining the extent of radioactive decay in ancient rock

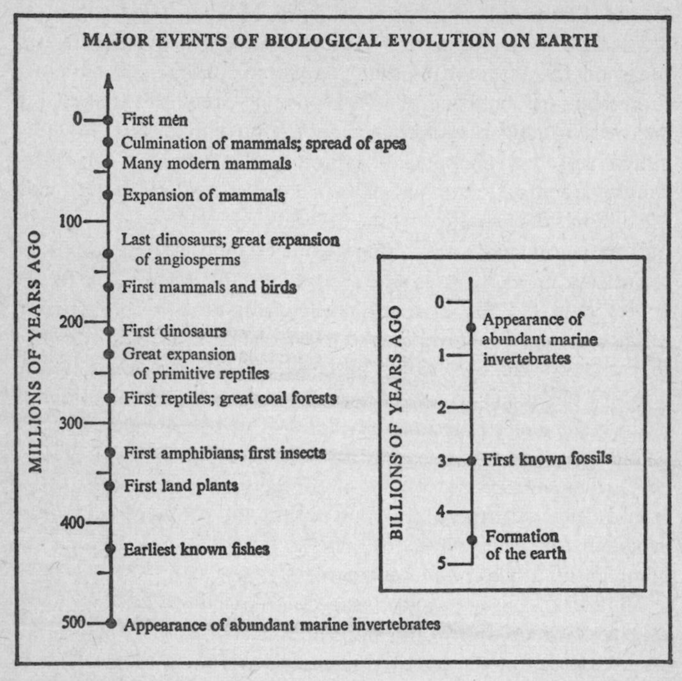

This chart was drawn by Huang to illustrate the brief portion of evolutionary history during which man has existed.

formation, it has been possible to date many events in the history of our planet and to set its age, in its present form, at some 4·5 billion years. It is thought that life emerged more than 2·5 billion years ago, leaving one or two billion years for the chemical evolution that culminated in the first living organisms. This is a far cry from the 100,000 years envisioned by Michelet a century ago.

There is a tendency today for scientists to dodge the issue of just what stage marked the first appearance of life. Because its characteristics evolved one by one, they say, trying to define the moment of life's 'birth' is like trying to agree when middle age begins.

Nevertheless, in June 1961 two dozen specialists from the

United States and abroad met at Big Meadow Lodge, on the Skyline Drive in Virginia, to discuss the geologic time scale as it bears on the question of life's emergence. Elso S. Barghoorn, Professor of Botany at Harvard University, reported that coal in northern Michigan had been assigned an age greater than 1·7 billion years and there was one report of patterns in limestone, like those left by algae, in a formation thought to be more than 2·7 billion years old.

Some participants felt it should be possible to date the 'oxygen revolution' that occurred when photosynthesis began and oxygen appeared in the atmosphere. The fuel that drives our cars, jet planes and industrial machines is largely, if not entirely, the fruit of that revolution, for the energy in coal and petroleum is primarily that which was once captured by plants from sunlight. The conferees, however, found themselves far from agreement on when the oxygen revolution occurred.

They also discussed the strange fact that the known fossil record does not, generally, go back beyond some 500 million years. It has been suggested that before then the planets and animals did not produce skeletons and shells hard enough to leave their mark in the sands of time. Yet, at the rather abrupt start of the fossil record, almost all of the primary divisions of the plant and animal kingdoms – the phyla – had evolved, and it seems hard to believe that by then they did not have fairly durable shells and skeletons.

Further radical revisions of the time scale, so far as the period available for chemical evolution is concerned, do not appear likely and, as stated earlier, it is this almost incomprehensible span of time that makes the emergence of life seem plausible. As one scientist has put it, 'In two billion years the impossible becomes the inevitable.'

Furthermore, as pointed out by Su-Shu Huang when he was still at Berkeley, since the process is controlled by the mathematical laws of probability, it is reasonable to suppose that the evolution of life on any other planet like our own would take a comparable period.

9

The Origin of Life: Building Molecules

IN contrasting Pasteur's time, when people like George Sand and Alexandre Dumas participated in the scientific debates, with the situation today, René Dubos has said:

> The great pageant of science is still unfolding; but now, hidden behind drawn curtains, it is without audience and understandable only to the players. At the stage door, a few talkative and misinformed charlatans sell to the public crude imitations of the great rites. The world is promised cheap miracles, but no longer participates in the glorious mysteries.

It is probably unfair to dismiss those at the door as charlatans, but otherwise what Dubos said is applicable to the current inquiry into the chemistry of life, in particular the mechanisms of heredity. Yet it is possible, in a general way, to grasp the steps whereby it is now thought the complex stuff of our bodies evolved from a few simple gases.

Four types of molecule were vital fruits of this process:

Proteins: These are among the most important structural materials of all living things and, in the form of enzymes, serve as the catalysts without which the chemical reactions of life would proceed at a snail's pace, if at all. A protein molecule consists of hundreds of amino acids linked in a chain that tends to form a spiral with hydrogen atoms as rungs, holding the spirals firmly in place. While some eighty amino acids are known, only about twenty figure in the construction of proteins. Like the twenty-six letters of the alphabet, they can be arranged in countless sequences that spell out their various functions.

Nucleic Acids: The substances responsible for the essential quality of life: continuity. The form known as DNA (deoxyribonucleic acid) remains in the nucleus of the cell as the repository of the instructions for the cell's functioning. Its cousin, RNA (ribonucleic acid), carries the instructions from the DNA to those portions of the cell that manufactire proteins. Amino acids hook together, during this process, to conform to the DNA pattern. DNA molecules are double spirals, arranged like a twisted ladder of great length. The sides of the ladder consist of sugar and phosphate units; the rungs are paired purines and pyrimidines (a typical one, adenine, is illustrated on p. 113). There are only four purines and pyrimidines in DNA: adenine, cytosine, guanine and thymine. They carry the messages like the dots and dashes of the Morse code. Those within RNA are the same, except that they include uracil instead of thymine.

Lipids: The fatty materials that store energy and form part of the cell structure. Their molecules consist of hydrogen atoms and some oxygen atoms mounted on a skeleton of linked carbon atoms.

Polysaccharides: The chains of sugar molecules that store energy (as in starch) and, in the form of cellulose, are a constituent of cell walls. A cellulose molecule consists of some 2,000 units of glucose. The polysaccharides are part of the larger family of carbohydrates.

It is noteworthy that the four prime elements in these vital substances – hydrogen, oxygen, nitrogen and carbon – are the four most common chemically active elements of the universe. While carbon, hydrogen and oxygen are common to all of these molecules, only the proteins and nucleic acids contain nitrogen. In addition, there is sulphur in many proteins, and phosphorus is an essential component of the nucleic acids.

By 1957 the volume of new information and new speculation on the evolution of these substances was such that it was thought advisable to hold an 'International Symposium on the Origin of Life on the Earth'. It was sponsored by the newly formed International Union of Biochemistry and was held in Moscow with Oparin and the Academy of Sciences of the U.S.S.R. as host. The site, however, did not imply any dominance by Marxists. In fact the more than forty specialists who gathered from sixteen countries represented a broad range of ideologies. Above all, by their distinction they demonstrated the emergence of this field of inquiry into full scientific respectability.

Among the Americans, largest of the visiting delegations, three men (Melvin Calvin, Linus Pauling and Wendell M. Stanley) had won, or were about to win, science's highest honour: the Nobel Prize. A fourth American Nobel laureate, Harold C. Urey, though unable to attend, contributed a paper and his former student at the University of Chicago, Stanley L. Miller, came to tell of experiments, inspired by Urey, that some considered the most significant to date, with regard to how life's building blocks were first synthesized.

The conference took up, in succession, the five steps necessary for the transformation of inanimate matter into living cells, namely:

1. The formation of the simplest organic compounds.
2. The transformation of these into more complex organic compounds.
3. The origin of the key life chemicals, such as proteins and nucleic acids.
4. The origin of structures and metabolism (energy-producing chemistry).
5. The evolution of metabolism.

The participants described experiments designed to test three basic ways whereby the simpler organic compounds might have been synthesized: by radiation, by lightning (electric discharges) or by heat. Of these studies, the first were those carried out by Calvin and his colleagues at the University of California in Berkeley during 1950.

Ever since the end of World War II Calvin's group had been studying the complex chemistry of photosynthesis, using carbon 14, the newly discovered radioactive form of carbon, as a tracer. In photosynthesis, plants manufacture carbohydrates and oxygen from carbon dioxide and water with light as their energy source. The plant studied by Calvin was Chlorella, a green alga. A known portion of the carbon in the carbon dioxide made available to the plant was radioactive and, by interrupting the process for analysis at various stages, it was possible to follow this carbon through the sequence of chemical reactions that finally led to synthesis of carbohydrates. The elegance of this work won Calvin his Nobel Prize, announced during the 1961 Green Bank conference on extraterrestrial life.

Following the photosynthesis studies, Calvin and his co-workers sought to explore whether or not high-energy particles from space (cosmic rays) or from radioactive substances within the earth's rocks might have played a role in synthesizing the original organic compounds. The intensity of such radiation from space and from the rocks is so slight that we are hardly aware of it, but it seemed possible that it could have been a significant source of energy over periods reckoned in millions of years. Furthermore, such radiation could have been more intense in the past.

The mixture initially bombarded by Calvin's group consisted of two gases that seemed to them probable components of the earth's primitive atmosphere: carbon dioxide and water vapour (in some runs ionized iron was also present). As in the photosynthesis studies, a known portion of the carbon dioxide had been tagged with radioactive carbon.

The bombardment was by a beam of alpha particles (the nuclei of helium atoms) accelerated to 40 million electron volts by the Berkeley cyclotron. The sample from each run was subjected to refined forms of analysis, such as paper chromatography, unavailable to researchers of earlier periods, and this disclosed that the bombardment had produced simple carbon compounds, in particular formic acid and formaldehyde (both molecules with a single carbon atom).

The next series of experiments – and to many the most critical during this early period – were those of Stanley Miller. As noted earlier, he was working at the University of Chicago under Harold Urey. The latter had pondered the low production rate of organic compounds in Calvin's experiment (apart from formaldehyde and formic acid) and suspected that the trouble might lie in the assumed nature of the earth's original atmosphere. Perhaps, instead of being rich in oxygen, as it is today, it was composed of hydrogen compounds with no free oxygen at all, as proposed earlier by Oparin.

In such an atmosphere the basic elements needed for the synthesis of organic compounds would be available in the following forms:

Oxygen: In water (H_2O)
Carbon: In methane (CH_4)
Nitrogen: In ammonia (NH_3)
Hydrogen: In hydrogen gas (H_2) as well as in the other gases.

During the 1930s it had been discovered that the atmospheres of Jupiter and Saturn are rich in methane and ammonia, and methane was later detected, as well, in the atmospheres of the other big planets: Uranus and Neptune. This gave strong support to the idea that the original atmosphere of the earth was of the methane-ammonia variety.

Urey had proposed, Miller told the Moscow conference, that either ultra-violet light or electric discharges (lightning) could have broken up molecules in such an atmosphere and allowed them to fall back together into more complex organic compounds. Because ultra-violet light of the appropriate wavelengths is very difficult to work with in the laboratory, being absorbed by glass and air, it was more practical to experiment with electric discharges.

Miller used a sealed system of flasks and tubes in which water vapour was obtained by boiling a reservoir flask. The vapour was mixed with methane, ammonia and hydrogen and was then subjected, in another part of the system, to a 60,000-volt high-frequency spark. The vapour that condensed was returned to the reservoir and boiled again. This was kept up for a full week.

Within a day the water in the reservoir flask had become pink, and by the end of the week it was deep red. Miller, like the experimenters in California, used paper chromatography for his analysis. Among the products were glycine, simplest of the amino acids, and alanine, another important amino acid, plus lactic acid, acetic acid, urea and a large amount of formic acid. All told, in one of his experiments, 15 per cent of the available carbon (in the methane) went into the production of organic compounds (2·1 per cent of the methane was converted to glycine). This strongly supported the view that such compounds must originally have been synthesized in an environment dominated by hydrogen,

THE MILLER EXPERIMENT

Components of the 'Primordial Atmosphere'

$H-C(H)(H)-H$	METHANE
$H-N(H)-H$	AMMONIA
$H-O-H$	WATER
$H-H$	HYDROGEN

Some Products of the Experiment

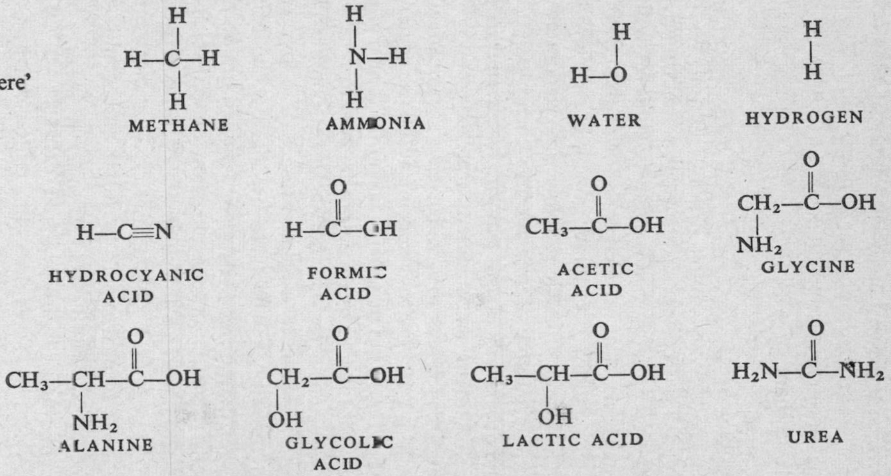

$H-C \equiv N$	HYDROCYANIC ACID
$H-C(=O)-OH$	FORMIC ACID
$CH_3-C(=O)-OH$	ACETIC ACID
$CH_2(NH_2)-C(=O)-OH$	GLYCINE
$CH_3-CH(NH_2)-C(=O)-OH$	ALANINE
$CH_2(OH)-C(=O)-OH$	GLYCOLIC ACID
$CH_3-CH(OH)-C(=O)-OH$	LACTIC ACID
$H_2N-C(=O)-NH_2$	UREA

The dashes between units of a molecule represent bonds. A hydrogen atom, for example, has one bond; a carbon atom has four. Two dashes represent a double bond; three dashes, a triple bond.

rather than oxygen. Furthermore, the high rate of production led Urey to suggest that the primitive ocean might soon have become rich in organic compounds, reaching a level perhaps as high as one per cent. Thus it was found that, in Miller's words, a hydrogen-dominated atmosphere was 'the key to synthesis of the organic compounds'.

During the Moscow discussions two Russians, T. E. Pavlovskaya and A. G. Passynsky, told of experiments confirming Miller's results. Others found that synthesis could be achieved with ultraviolet light (particularly at wavelengths shorter than 2,000 angstroms) and with gamma rays. Sidney W. Fox of Florida State University, who was at the Moscow meeting, generated amino acids by heating a mixture of urea and malic acid to 300°. However, Urey and others later questioned whether, even if there were an abundance of hot springs, such temperatures could have been widespread enough for heat to have played a significant role.

Meanwhile Philip H. Abelson, working at the Carnegie Institution of Washington, found that amino acids could be synthesized, in the manner of Miller's experiment, using a wide variety of gas mixtures. It was only necessary that the gases contain the four basic elements (carbon, hydrogen, nitrogen and oxygen) and that their composition be such that hydrogen could exist in a free state, rather than oxygen. He also identified amino acids in fossils as much as 300 million years old, showing that at least some amino acids are stable for long periods of time provided the temperature is low enough (80° or lower). This means that the amino acids, once formed, would be stable and would accumulate in the oceans of the primitive earth.

The apparent role of hydrocyanic acid (HCN) in the Miller experiment led others to explore this matter further. It was suspected that Miller's sparking of primordial gases first produced this acid, which then reacted with the other substances. At the University of Houston, in Texas, a group led by Joan Oró, a Spanish-born biochemist, worked with water-based solutions of ammonium cyanide (NH_4CN) that generated amino acids, chains of amino acids and purines, in particular adenine, one of the message-carrying units of DNA, the most important of all biochemicals in that it transmits heredity.

Oró also made the startling suggestion that a considerable portion of the basic compounds came from comets. In 1908 an explosion more fearsome than that of an outsized hydrogen bomb occurred in the air over Central Siberia. Apparently the head of a comet had plunged into the atmosphere (see the account on page 130). From studying the glow of comets and their tails as they near the sun, we know they contain carbon compounds, particularly cyanide, and Oró suggested that, during the first two billion years of the earth's history – the critical period of chemical evolution – enough comets hit the earth to contribute between 200 million and 1,000 billion tons of material. More of it may have been picked up from near misses and comet tails.

In Berkeley, Calvin, continuing his experimentation with particle bombardment, decided to explore its effect on a primordial atmosphere similar to that used by Miller. He assumed that the chief source of such particles in the distant past must have been radiation from potassium 40 in the earth's crust. This substance emits beta rays (high energy electrons), and Calvin simulated this by using a beam of electrons accelerated to five million electrons volts. As with Oró's experiments, one of the many organic compounds produced by the bombardment was adenine. The others included sugars with four, five, six and even seven carbon atoms in their molecules. In these experiments methane containing radioactive carbon was used, so that there would be no doubt as to the source of the carbon in the resulting compounds.

Despite the success of these efforts, Calvin finally concluded that ultra-violet light was probably the chief agent in synthesizing the original organic compounds, with radioactivity and electric discharges as secondary contributors. He felt that radiation from space – the so-called cosmic rays – was a relatively unimportant source of energy.

By the early 1960s evolutionary paths for the building blocks seemed clear. In particular several components of the most important molecules of all – those of DNA and RNA that maintain the continuity of life – had been synthesized in simple ways. The two sugars that are bound into the skeletons of these molecules (ribose and deoxyribose) were among those that had been produced merely by shining ultra-violet light onto a formaldehyde solution. The latter, in turn, had been generated by various of the experiments with a primordial atmosphere. At least three of the five pyrimidines and purines that spell out the messages in DNA and RNA had been synthesized (adenine, guanine and uracil), although, in the case of uracil, some considered the method of synthesis unrealistic for the primitive earth.

Most exciting was the realization that, out of all the possible combinations of atoms, the ones most readily synthesized in the laboratory from components of the earth's assumed primitive environment were generally those which predominate in the huge molecules of life – the key amino acids, sugars, purines, pyrimidines and fatty acids.

The next problem was to figure out how the huge, composite molecules, such as DNA, RNA, the proteins and the catalysts that make possible life's chemistry, were put together. Calvin believes that a phenomenon known as autocatalysis helped produce the catalysts themselves. In this process the product of a reaction acts as the stimulus (or catalyst) to that same reaction. This promotes that particular reaction at the expense of rival processes and should lead to a form of natural selection of certain catalytic substances in the course of chemical evolution. Such a process, for example, could have produced the modern substance, catalase, starting in the primitive ocean with hydrogen peroxide (H_2O_2) and ferric ions. The latter are iron atoms that lack three electrons and hence have a triple positive charge. They are found in seawater and serve as a crude catalyst in transformations of hydrogen peroxide. Some of these reactions produce far more efficient catalysts that tend to monopolize the hydrogen peroxide reactions, but are, in turn, susperseded by other even more efficient catalysts. The catalase that finally emerges, according to Calvin, is 10 billion times more efficient than the original ferric ions.

In this manner, he believes, such important catalysts as the iron porphyrins evolved. They resemble the chlorophyll that enables plants to carry out photosynthesis and, apparently, it was random variation in the construction of the porphyrins that led to the appearance of chlorophyll. The haemoglobin that carries oxygen in our blood and gives it its red colour is related to chlorophyll, testifying to our common ancestry with the plants.

Among those who sought to explain how the building blocks hooked up like elongated freight trains to form long-chain molecules was Sidney Fox, who had argued in favour of heat as the agent that synthesized the amino acids. He turned to the same stimulus for 'polymerization' of these giant molecules. He took the eighteen amino acids that are found in almost all proteins and, so to speak, put them in a pot and cooked them together. He defined what emerged as 'proteinoid' – a substance that resembled protein in that bacteria ate it and enzymes digested it. More startling, when dissolved in hot water and allowed to cool, it formed into a multitude of microscopic spheres – some 50 billion of them per gram – that looked very much like cocci, bacteria of the type that cause pneumonia. If polyphosphoric acid was added to the solution, it was no longer necessary to heat it above the boiling point of water (shades of Darwin's 'warm little pond' rich in ammonia and phosphoric acid salts!).

Fox did not claim he had manufactured living bacteria, but he felt his experiments had shown that heat could have played the key role in building up the giant molecules.

Some of the most recent experiments have been directed to possible paths whereby one of the most vital life chemicals may have

THE STRUCTURE OF ATP

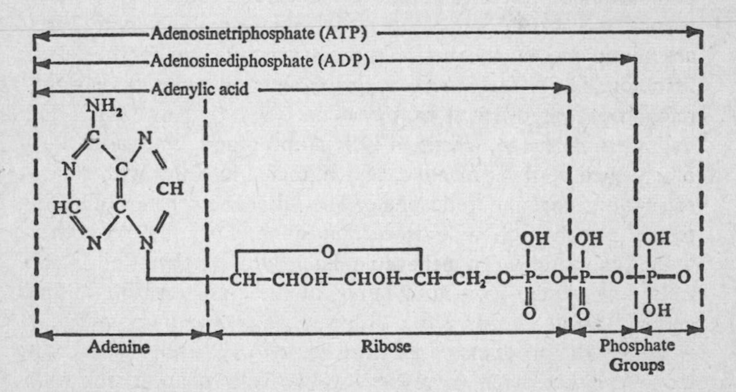

Heavy dashes indicate the energy-storage bonds between phosphate groups.

evolved. This substance, ATP (adenosine triphosphate), delivers the energy that enables life to perform its mechanical functions, be they within the most primitive one-celled organism or in man. ATP is thus the 'energy currency' of the body. No muscle could move without it.

The simplest chemical in the ATP family is adenylic acid (adenosine monophosphate). A molecule of this substance consists of one unit each of adenine, ribose sugar and phosphate. When an added phosphate is hooked on, the substance becomes adenosine diphosphate and when a third phosphate is added, it is adenosine triphosphate – ATP. The energy required to make these bonds is money in the bank, so to speak. ATP picks up such energy, in photosynthesis or in the cell's 'Federal Reserve' of energy, by hooking on a phosphate group. It delivers the energy, where needed, by dropping such a group. ATP also appears to be a key building block in RNA synthesis and differs in only one atom from a similar stage in DNA formation.

Efforts to produce ATP under conditions of the primitive earth have been carried out by a trio that included two researchers at the Exobiology Division of NASA's Ames Research Center: Cyril Ponnamperuma of Ceylon and Ruth Mariner. Both had worked with Calvin on experiments that, through electron bombardment, generated rich amounts of adenine. The third participant, Carl Sagan, then at Stanford University, was becoming one of the nation's most versatile and enthusiastic 'exobiologists'. The term, a newcomer to scientific jargon, refers to extraterrestrial biology.

Following the technique of Calvin and others, they used adenine tagged with radioactive carbon, mixed in water with ribose sugar and ethyl metaphosphate. The latter is a cluster of phosphates attached to a hydrocarbon spine. For one hour they bathed the solution in ultra-violet light (at a wavelength of some 2,500 angstroms) from four lamps of the kind used for killing germs.

To test the products of the presence of ATP they used a technique that had been developed by two researchers at the Oak Ridge National Laboratory in Tennessee. It exploited the discovery that ATP causes the lanterns of fireflies to glow. A large number of dehydrated fireflies tails were obtained from Schwarz Bioresearch, Inc., of Mount Vernon, New York, a firm that specializes in such unusual requirements. Luminescent material from the tails did, indeed, show that considerable quantities of ATP had been synthesized.

The researchers pointed out that ATP is a key product of photosynthesis in today's plants, but the synthesis is no longer performed under the influence of ultra-violet because such light, at the proper wavelengths, cannot penetrate the ozone in the contemporary atmosphere. Early in the history of life, they said, photosynthesis of ATP took place outside of living organisms, with the help of ultra-violet light. Hence ATP and its energy were everywhere to be had, like the oxygen of our air. However, during this period water vapour high in the atmosphere was being broken up by sunlight into its components, hydrogen and oxygen. The earth's gravity was too weak to hold the light-hearted hydrogen atoms, so they gradually escaped into space, and oxygen began to appear as a stable component of the air. Enough of it formed ozone (O_3) to cut off the ultra-violet, and the synthesis of ATP, without help from living organisms, began to fade. It was then that plants were forced to 'discover' the value of chlorophyll in synthesizing ATP and carbohydrates, using the energy of sunlight as it reaches the earth's surface today. Once they had done so, mass production of oxygen by the plants began. The 'oxygen revolution' which transformed the chemistry of the atmosphere and of the surface rocks was under way and, thanks to that revolution, oxygen-breathing life is possible on earth today.

By the mid-1960s a variety of long-chain molecules had been synthesized, although little light had been shed on how nature first produced the lipids and nucleic acids. However, the role of ATP and other substances of its class (polyphosphate esters) in helping to synthesize long-chain molecules within living cells led a group of Germans to explore the possible synthesizing influence they may have had before the appearance of life. The group, at the Max Planck Institute for Virus Research at Tübingen in West Germany, was headed by G. Schramm, who had taken part in the Moscow conference. They found that, with such polyphosphate esters, they could synthesize 'protein-like' chains of up to 24 amino acids; polysaccharides were similarly formed from sugar molecules and, by a three-stage process, long molecules somewhat similar to the nucleic acids were obtained. Just how this came about, chemically, is not clear at this writing, and the fact that the process took place in the absence of water pointed up its artificiality, for it is hard to see how nucleic acids could have been formed in nature except in water solution.

Nevertheless the German experiments, like those of Fox, resulted in strikingly lifelike structures. The molecules of polyadenylic acid that they produced were so long that, when magnified 60,000 times in an electron microscope, they strongly resembled the elongated fibres of RNA. 'Taken together,' the Germans reported, 'our experiments have shown that the most important macromolecules [giant molecules] can be prepared in simple fashion with the aid of polyphosphate esters.' Such long-chain molecules, they added, could not have evolved randomly, even in several billion years. In effect they refuted the simile of the typing monkey that ultimately writes *Hamlet*.

'Such an event can hardly have occurred within the finite time that the earth has existed,' they said. The long molecules must have developed through some systematic process, gradually improving their effectiveness, link by link.

Despite the progress that has been made in recent years, only the initial steps in life's evolution have been elucidated in a convincing way. The discussion so far has concerned the substances within the basic unit of life: the cell. But these chemicals interact and function in a manner so complex that it still has not been fully deciphered. We cannot understand the cell, and how it evolved, simply in terms of its constituents. As Harold F. Blum, Visiting Professor of Biology at Princeton, has put it: 'Clearly we should not try to describe an automobile by grinding up its various parts and subjecting them to chemical analysis, and we would not expect to learn all about the living machine by following, exclusively, a similar attack.'

In the December 1961 issue of the *American Scientist* he criticized the Moscow conference on the origin of life as having been preoccupied with the first stages of biochemical evolution without really facing the problem of how the substances of life organized themselves into the wonderful mechanism of the living cell. This theme was picked up in 1964 by George Gaylord Simpson, Professor of Vertebrate Paleontology at Harvard and a specialist in theories of evolution.

The task of reconstructing how life and its early forms did, in fact, emerge is made enormously more difficult by our lack of any fossil records of that time. Apart from early traces of algae, the earliest usable record begins only a half billion years ago, after the major decisions of evolution already had been made. In fact the entire geological history of the earth's first billion years seems to have been ploughed under or otherwise lost.

Blum and Simpson are chief dissenters against the optimism of those who believe life will arise on any planet like the primitive earth. Simpson concedes that the progression from individual atoms to the long-chain molecules is probable – perhaps even inevitable, on such a planet. However, following the arguments of Blum, he says the next jump – to a living cell – is so vast that it must be comparatively rare.

The problem, like that of the typing monkey, is one of probabilities. As Calvin has pointed out, the early stages of chemical evolution seem highly deterministic. One can assume that the same thing happens on any earthlike planet. Even the steps leading to the first live cell, Calvin believes, are controlled by chemical and physical laws. A vast number of chemical combinations are given a chance to qualify for what emerges as life, but the outcome is preordained. However, as evolution proceeds and its fruits become more complex, the number of alternate paths also increases and the outcome becomes less and less predictable. Hence Calvin says that one-celled life on another planet might resemble its counterparts on earth, but the higher forms that followed would differ radically.

The Blum-Simpson argument is simply that diversity of paths comes much earlier in the game – before any life appears. Life is but one of many possible outcomes, they say. The disagreement does not concern the existence of extraterrestrial life, but the question of its frequency of appearance – and hence its proximity to the solar system. In challenging the avowed goal of the American space programme to seek out extraterrestrial life, Simpson said: 'There probably are forms of life on other planetary systems somewhere in the universe, but if so it is unlikely that we can learn anything whatever about them, even as to the bare fact of their real existence.' Indignant at the cost of the American space programme, Simpson conceded that his arguments were 'subjective and speculative', but so, he added, were those of the biochemists and others who argue hopefully that life is inevitable. The truth is that we still do not know enough about the process of evolution to resolve the debate. The answer may be that, if paths alternative to that which led to the first cell were almost unlimited, as implied by Blum and Simpson, life would probably not have arisen even on earth. There must, therefore, have been some guidelines to steer biochemical evolution in the right directions. Calvin's concept of autocatalysis may be but a simple example of far more complex trial-and-error processes that were involved.

Miller, too, takes issue with the views of Blum and Simpson. The evolution of cellular organization and function looks difficult, he says, because we do not understand it. Yet a generation ago the same was true of chemical evolution. The latter process no longer seems improbable because its feasibility has been demonstrated. He suspects that the time is not distant when the subsequent steps also will be understood and appear straightforward. He objects to the heavy emphasis laid by some on the need for great periods of time – the argument that, given enough time, almost anything can happen. The demonstrated characteristic of chemical evolution, whereby it largely follows predetermined paths, may, contrary to the argument of Blum, extend farther along the highway to life than we can now explain. If so, life would arise rather early in the history of an earthlike planet.

The question is often raised as to whether or not life on another planet could have evolved with a chemistry based on substances other than carbon and water. Urey and Miller, in a 1959 analysis of the problem, said flatly: No.

'We know enough about the chemistry of other systems,' they wrote, 'such as those of silicon, ammonia and hydrogen fluoride, to realize that no highly complex system of chemical reactions similar to that which we call 'living' would be possible in such media.'

For example, it has been argued that giant molecules built on chains of silicon atoms, instead of carbon atoms, would lack the side chains that play an essential role in life's chemistry. It has also been pointed out that ammonia is a liquid only within a very narrow and frigid temperature range, making it a poor substitute for water as a medium in which life could evolve and flourish. Another perhaps critical advantage that water has over ammonia is a special property that it displays when it freezes: it expands. Ice is therefore less dense than water and floats. Were it otherwise, there might be no life anywhere. Every chunk of new ice formed on the surface of the lakes and oceans would sink to the bottom and stay there. The water regions of every planet, according to George Wald, Professor of Biology at Harvard, would become frozen solid and would not thaw, even in prolonged periods of warm weather. This realization led him to wonder what would happen in an ocean of ammonia. Through experimentation he found that frozen ammonia sinks in liquid ammonia (he later discovered that the experiment had been performed 20 years earlier). He concluded that life could not arise in an ammonia ocean.

In any case the consensus of virtually all who have studied the problem is that life exists elsewhere than on earth. As Calvin pointed out, we cannot expect the process of evolution, elsewhere, to be at our particular stage of development. The emergence of man has taken only about a million years of a total evolutionary span of several billions of years. Hence evolution on other earth-like planets must in many cases, be pre-cellular. In other cases it may be 'posthuman', to use Calvin's word. The latter instances would, of course, represent more advanced stages of evolution than our own. The former – if we find such material, for example, on Mars – could help us understand that part of life's history that has been wiped out on earth.

We can assert with some degree of scientific confidence [Calvin concluded] that cellular life as we know it on the surface of the earth does exist in some millions of other sites in the universe. This does not deny the possibility of the existence of still other forms of matter which might be called living which are foreign to our present experience. ... We have now removed life from the limited place it occupied a moment ago, as a rather special and unique event ... to a state of matter widely distributed throughout the universe.

These words were written in 1958, one year after Calvin sat as chairman of the closing session of the Moscow conference and one year before the initiation of Project Ozma, the first serious effort to search for signals from more advanced creatures than ourselves.

10

Visitors from Space

AT approximately 5.30 p.m. on 15 March 1806 a man named Reboul and his son Mazel, employees of a local landowner, were working in the fields near the village of Valence, in the south of France, when they heard the unlikely sound of cannon shots. Napoleon's armies were far to the east, for the battle of Austerlitz had been fought only three months earlier. Furthermore, the sound seemed to come from the wintry sky and was followed by prolonged thundering of a peculiar sort. This, in turn, was succeeded by a sound that, the two men said later, resembled the scream of a spinning well hoist that has been released, allowing the bucket to plunge downward.

A moment later they were transfixed by the sight of an object flying through the air toward them. It plunged into the ground fifteen paces from Reboul, who cautiously drew near. He found a lump of black material, about the size of a child's head, that had broken into three pieces.

Meanwhile, a few miles away, another father-son team in the fields had a similar experience. They told the scientifically inclined burghers who came to investigate from the nearby town of Alais that they saw a dark body come flying out of the clouds and land near them, bursting into fragments and digging a shallow pit.

Reboul and his son were persuaded to part with one of their three chunks, and another was obtained from the second batch of fragments. The rest have been lost to science – a pity, for this proved to be the first known fall of a carbonaceous chondrite, an extremely rare and interesting visitor from heaven. It is known, because of its proximity to that town, as the Alais meteorite.

Some twenty-eight years later a fragment of Alais reached J. Jakob Berzelius, the great Swedish chemist, who took one look at it and decided there must have been some mistake. This could not be a meteorite. It was too crumbly, and the material disintegrated in water. There were then thought to be three kinds of meteorite, all of them hard and insoluble: those of iron (with an admixture of nickel), those of stone, and those of iron filled with bits of stone like a raisin cake.

Berzelius was about to throw the specimen away when he re-read the description of the object as it was found, fresh and warm after the fall. This, and the evidence that a paper-thin layer of the original surface had been fused by heat, convinced him that the object on his desk had really fallen from the sky. He set to work, with the best methods of analysis available in 1834, to determine its composition.

To his amazement, it was rich in carbon compounds and responded to his tests much as would humus, the mixture of decayed vegetable and animal matter that enriches our soil. Might it, in fact, be genuine humus or some other assemblage of organic compounds, he asked in his report. 'Does it possibly give an indication of the presence of organisms on extraterrestrial bodies?'

His question is still unresolved after the passage of 130 years. In fact, with the application of new analytic methods, it has become one of the more hotly debated problems of modern science.

Berzelius himself decided in the negative. Despite the resemblance of carbon compounds in the meteorite to those in the soil of our own planet, he said, this 'does not appear to justify the presence of organisms in its original source.' He felt that rock on the parent body whence came this 'chunk of soil' must have been converted into earth by some unknown process.

Four years after Berzelius did his analysis another of these strange objects was seen to fall, this time near the Cold Bokkeveld Mountains in South Africa. In 1857 a third was recovered at Kaba, near Debrecen, Hungary, and samples of Kaba and Cold Bokkeveld were sent to Berzelius's most famous student, Friedrich Wöhler at Göttingen, Germany. It was Wöhler who, some years earlier, had achieved the first laboratory synthesis of an organic compound (urea). From Cold Bokkeveld he extracted an oil 'with a strong bituminous odour'. Although the proportion of organic material was very small, it was definitely there, he said, and, after his analysis of both specimens, he concluded that, 'according to our present knowledge', such substances could only have been produced by living organisms.

That was in 1860. Four years later, on the night of 14 May1864, the peasantry in the south of France had another terrifying experience. An object as large as the full moon, but more like the sun in brilliance and slightly teardrop-shaped, raced across the sky with the sound of an express train, punctuated by peals of thunder. It was visible throughout Aquitaine as it broke into pieces that quickly darkened. Behind it a broad, luminous trail turned to white smoke and slowly dissipated. When dawn came, it was found that parts of the meteorite had rained on the region around the village of Orgueil. Scientists from the town of Montauban recovered twenty pieces, some head-sized, but most of them smaller than a fist. They found that the specimens could be cut with a knife and, when sharpened, could be used like a pencil.

A specimen of this, the Orgueil meteorite, was immediately examined by a French scientist, S. Cloëz, who found it was held together by a water-soluble salt. When moistened, the material immediately turned to dust. In fact, one chunk of this meteorite, preserved in the humid environment of a museum in Calcutta, India, has disintegrated into dust spontaneously. Cloëz was much impressed by the resemblance of the carbon, hydrogen and oxygen content of his sample with that found in peat or lignite (a woody variety of coal). Such substances in this and other meteorites, like Alais, he said, 'would seem to indicate the existence of organized substances in celestial bodies.'

By 'organized substances' he meant life. His report was communicated to the Academy of Sciences in Paris by Gabriel-Auguste Daubrée, one of the six members of its Mineralogical Section, another of whom was Louis Pasteur. It was, in fact, the year of Pasteur's great lecture on spontaneous generation. Pasteur said it could not occur on earth, yet here was possible evidence it had taken place elsewhere. The report of Cloëz, coming at the height of this controversy, must have created a sensation, which may account for a strange discovery concerning the Orgueil meteorite that, as will be seen, was not made until a century later.

All told, by 1964, only 20 meteorites of this kind had been identified out of a total of more than 1,500 well-authenticated falls and finds. A 'fall', as opposed to a 'find', is a specimen whose arrival on earth is witnessed. Many falls produce great numbers of fragments. One that struck Pultusk, Poland, in 1868 is said to have broken into at least 100,000 stones, and the Sikhote-Alin meteorite, which fell on the mountains of that name flanking the east coast of Siberia, showered the region with iron on 12 February 1947. At least 60,000 pounds of it have been collected.

While iron meteorites predominate among those found, because they are so distinctive, stones constitute 92 per cent of the observed falls, with irons responsible for only 6 per cent. The remaining 2 per cent are of the intermediate 'stony irons'. Stones dominate the observed falls partly because, in space, they break up more easily than irons. For a true assessment of what hits the earth one should take bulk into account and here the irons may predominate, if only because the fragments of one such fall, Sikhote-Alin, tip the scale heavily in their favour.

The 20 known falls of the peculiar, carbon-rich meteorites are included among the stones, despite their crumbly consistency. The most recent of this type known to have entered the earth's atmosphere brought itself to the attention of a large part of the population in the Dallas–Fort Worth area of Texas on 9 September 1961. It was a Saturday night and the drive-in movies were full. Suddenly, at some of them, the picture vanished in the glare of a blinding light, as though the sun had suddenly been turned on. Those who looked up in time could see a fireball flying overhead.

The Miss America contest was on and Miss Texas was a leading contender. Many were tuned in to KRLD, a local television station on which the contest was being shown. To their dismay, as one man put it, 'the television picture went all to pieces.' Thousands of viewers got to their feet and tinkered with the controls, unaware that the real trouble was in the sky overhead. The meteorite had left a trail of ionization that interrupted the TV signals.

Brian Mason, Curator of Mineralogy at the American Museum

of Natural History in New York, believes the original weight of the object may have run into tons, but only a half dozen fragments, totalling about 11 ounces, were found along a four-mile stretch of land near Bells, Texas. The trouble was that Hurricane Carla drenched the area soon after the fall, perhaps disintegrating some of the specimens.

Their material looked so much like dirt that it could have escaped the attention of even the most experienced meteorite-hunter. This, plus the vulnerability of such specimens to weathering, has limited the finds to those whose fall was observed. Some meteorites in our museums fell in prehistoric times, but not those of this variety.

Because these meteorites, known as carbonaceous chondrites, have been alleged to show that life had existed elsewhere than on earth, their origin and that of meteorites in general merits attention. The term carbonaceous chondrite derives from the fact that these specimens are rich in carbon and permeated with rounded beads known as chondrules. Roughly fourteen out of every fifteen stony meteorites contain chondrules. Nothing like these beads of magnesium-iron silicate has been found in terrestrial rocks. For generations scientists have speculated as to their meaning, with regard to the origin of meteorites. In fact, it was not until Jean Baptiste Biot, the French physicist, actually witnessed a fall of some 2,000 stones in 1803 that the scientific world fully accepted the idea that they come from the sky.

Yet meteorites have been known – and even worshipped – for thousands of years. The Black Stone in the Ka'ba at Mecca is the holiest of holies in the Moslem world, having been kissed by countless pilgrims since before the days of Mohammed. To free his followers from idolatry, Mohammed destroyed various sacred objects in the Ka'ba, but he spared the Black Stone. In the excavation of a Mexican temple a meteorite was found, wrapped in mummy cloths. Another, in India, was long anointed daily and decorated with flowers.

Even before the Iron Age meteorites provided material from which knives were made – weapons valued at many times their weight in gold. The largest meteorite on exhibit is a 31-ton chunk of nickel-iron brought back from Cape York, Greenland, by Robert E. Peary, discoverer of the North Pole. It is at the American Museum of Natural History in New York. Meteorites, as they exist in space, seem to come in all sizes, from chunks miles in diameter down to dust grains.

It is calculated that any meteorite heavier than 100 tons hits the earth with such force that a major portion of it is vaporized by the resulting heat. In other words, it blows up, producing a characteristic crater with raised rim. The classic example of such a feature is Meteor Crater in Arizona, whose width is three-quarters of a mile. It was produced by an object estimated to have weighed from 100,000 to one million tons that hit the earth at 35,000 miles an hour. All but one per cent of it presumably vaporized in the resulting explosion. The huge craters of the moon suggest that, on occasion, both the moon and earth have been struck by very large objects. Harold Urey has proposed that a considerable portion of the lunar craters were produced during the earliest period of the earth's history, when space was still cluttered with large objects and things had not yet settled down. Nevertheless, it is obvious that such impacts have been continuing. On the earth erosion and sedimentation, both by-products of our weather, have erased the evidence of most ancient features, as has the repeated ploughing under of the surface rocks by the forces that built our mountains.

In spite of these influences there are hints of giant craters in the earth's landscape, although until recently there was no way to tell whether or not they were produced by meteorites. The story of how this question was settled goes back to 1891, when a tiny diamond was found in one of the meteorite fragments picked up near Meteor Crater. The fragment is now at the Smithsonian Institution's Museum of Natural History in Washington, D.C. Diamond is a form of carbon created at depths of 200 miles or more within the earth, which is why it is so rarely found near the surface. Whereas graphite is an arrangement of carbon atoms that can be produced under normal pressure, it takes the weight of hundreds of miles of rock to squeeze carbon atoms into the diamond configuration. It was this, in part, that led Urey to presume that the diamond-bearing meteorites must have been formed inside of bodies at least as large as the moon.

In November 1959 the National Aeronautics and Space Administration called a conference in Washington on the origin of the moon and meteorites. Among those invited were two members of the faculty of the University of Chicago: Edward Anders of the Enrico Fermi Institute for Nuclear Studies and John C. Jamieson of the Geology Department. Neither knew the other, but they recognized names, as they registered at the Willard Hotel, and they went to the bar for a drink. Anders had just been given some diamonds from the debris around Meteor Crater and the conversation turned to their origin. Anders and his student, Michael E. Lipschutz, had been wondering whether it was possible that these crystals had been produced by the shock of impact, rather than by prolonged pressure. This was not a new idea. People long before had wondered whether they could achieve sudden wealth by pounding carbon, but it never worked. However, it occurred to the two men that there is one kind of graphite – the rhombohedral form – whose atoms are arranged in the same manner as in diamond, except they are slightly farther apart. Might it be possible, by hitting this graphite hard enough, to produce synthetic diamonds? Jamieson, by a fortunate coincidence, had been doing some experiments on the effects of shock with Paul De Carli at the Stanford Research Institute in California. The idea was put to the test in January 1960 and, sure enough, tiny diamonds were produced.

Soon after this was reported in the journal *Science*, letters began arriving from the big industrial diamond concerns – de Beers, Krupp and others – asking for reprints. Obviously they were tantalized at the thought of 'instant diamonds'. Meanwhile a number of concerns had been moving toward the high-pressure synthesis of diamonds, an important step being the development of equipment for such work by Loring Coes, Jr, of the Norton Company in Worcester, Massachusetts. In 1953 Coes was able to squeeze quartz into a dense state never before observed. The new substance, like quartz, was still silicon dioxide, the prime constituent of glass. Its chemical composition was unchanged, but its atoms had been compressed into a tighter configuration.

In honour of Coes the new substance was named coesite. In their experiments in transformation by shock De Carli and Jamieson tried to pound quartz into coesite without success. Eugene M. Shoemaker of the United States Geological Survey looked in vain for coesite in quartz-bearing material subjected to a nuclear explosion.

Then Shoemaker and two colleagues, Edward C. T. Chao and B. M. Madsen, decided to look at the fragments of Coconino sandstone from in and around Meteor Crater. Shattered pieces of this white quartz-bearing sandstone constitute a large part of the debris that lies, to a depth of 600 feet, under the crater floor. Samples from the crater rim and from a hole drilled 650 feet into the crater floor were analysed by X-ray diffraction, optical properties and spectrograph. All three methods disclosed the presence of coesite – the first time it had been seen outside the laboratory.

When Shoemaker, Chao and Madsen reported their discovery, in the 22 July 1960 issue of *Science*, they noted that coesite 'may afford a criterion for the recognition of other impact craters on the earth and perhaps ultimately on the moon and other planets.'

There was still another substance in the Meteor Crater debris that Chao and his co-workers could not identify. When X-rays were shone through the tiny grains of this material they were diffracted into a pattern unlike that of any known material. The following year S. M. Stishov, who was doing high-pressure research at Moscow State University, sent to his American colleagues a report on his experiments in the further compression of silicon dioxide. He said that with a pressure of 2,350,000 pounds per square inch and a temperature above 2,200 degrees he had squeezed this substance into an even more compressed phase than coesite. In the lattice structure of its crystals six oxygen atoms, instead of the usual four, were equidistant from each silicon atom. The new material was named stishovite, for its synthesizer.

On a computer Chao calculated the diffraction pattern to be expected when X-rays passed through stishovite crystals, and it proved identical with the unidentified pattern from the Meteor Crater debris. It probably takes the weight of 300 miles of rock to make stishovite, compared to only 200 miles for diamonds. The

Meteor Crater explosion must have been a terrible event indeed.

With such tags for the identification of impact craters, it was now possible to find out whether or not even large depressions were produced in this way. Coesite was soon found at several of them, including the Ries Kessel, a bowl 15 miles wide in western Bavaria. It has thus become evident that, from time to time, the earth is struck by huge meteorites that explode with violence greater than that of the most terrible nuclear weapon. Probably 1,000 or more meteorites hit the earth every year, most of them unobserved because they land in the sea or in remote areas. They tend to fall on the afternoon side of the earth, overtaking our planet in a manner that suggests they were in orbit farther out in the solar system.

Only one person is known to have been hit by a meteorite. On 30 November 1954 Mrs E. H. Hodges of Sylacauga, Alabama, was sitting in her home after lunch when a nine-pound stone crashed through the roof and hit her in the upper thigh. She was not seriously hurt. A few other hits have been reported but not authenticated.

While falls of extremely large meteorites are rare, the rate seems to increase steadily with smaller sizes until, at the microscopic level, there is a constant rain of material. On the basis of observations with rockets, satellites and deep space probes it has been estimated that this dust falls at a rate of about 1,000 tons a day, but the figure is still the subject of debate.

Meteors, as opposed to meteorites, are shooting stars – tiny points of light that streak across a segment of sky, then burn out. They are generally bits of frothy material that enter the atmosphere at high speed and disintegrate before penetrating very far. They differ from meteorites in being too insubstantial to survive a plunge through the atmosphere to the earth's surface. Almost all meteors seem to be debris left behind by comets, many of them still orbiting the sun along the highly elliptical paths of their parent comet. Such meteors are chiefly seen on those days each year when the earth crosses one of these paths. Among the better known of these 'meteor showers' are the Lyrids, which fall on or about 20 April, the Perseids of around 10 August, the Leonids of mid-November and the Geminids of 10–13 December. The fact that meteorites do not fall any more often during these showers than at other times has been thought to indicate that the two phenomena are unrelated. Thus the icy comets and their offspring seem to be an entirely different breed from the planets, asteroids and meteorites.

Do comets themselves ever strike the earth? Apparently they do. On 30 June 1908 a fireball streaked across the Yenisei basin in central Siberia, and from Kirensk, 250 miles away, a 'pillar of fire' was seen over the horizon, followed by several detonations. Gusts of wind swept across the steppes, so powerful that horses were knocked down near Kansk, more than 400 miles away. Earthquake recorders from Washington to Java detected the explosion and shock waves in the atmosphere circled the world in all directions, thus being recorded twice at Potsdam, Germany. For several nights thereafter the sky was strangely bright, from Spain to the Arctic Ocean, because of particles high in the atmosphere.

It was not until 1921 that the first scientific expedition reached the remote site, yet it was possible to locate the area of the explosion because trees had been felled outward in all directions for 20 or more miles. On hilltops they were levelled more than 35 miles away, their bark and branches often stripped off. At closer ranges they were charred. Yet no large crater or fragments were found – only microscopic bits of nickel-iron in the soil. Apparently this, the Tunguska 'meteorite', weighing several thousand tons according to the Russian estimate, exploded in mid air like a hydrogen bomb.

While some fanciful Russians have proposed that this was a nuclear device or spaceship sent from another world, the general feeling is that it was a comet head. It has led some to believe that, all told, the earth gains more material from comet impacts than from meteorites – a finding of special interest to those who believe comets were a primary source of carbon compounds essential to the emergence of life.

For a long time it has been thought that the meteorites are children of the asteroid belt. In this traditional view the asteroids orbiting the sun between Mars and Jupiter are the remains of a planet and the meteorites are smaller fragments thrown, perhaps by periodic collisions and near misses, into lopsided orbits that carry them close to the earth. The nickel-iron meteorites would be remains of the core; the stony meteorites would be from the mantle enveloping the core; the stony irons would be from the transition zone between core and mantle; the carbonaceous chondrites, with their hints of life, would be from the crust, or surface rocks. We suspect that the core of our own planet is of nickel-iron with an overlying mantle, some 1,800 miles thick, that may consist of magnesium-iron silicates such as the olivine and pyroxene found in stony meteorites. The differentiation of the earth into a core of very heavy material, a mantle of moderately heavy rock and a comparatively light crust is thought to have been brought about chiefly by heat generated within the planet by the pressure of its own weight and by its internal radioactivity. Now much of this energy has spent itself, but when the earth was young it was almost certainly much hotter; volcanoes were widespread; earthquakes and mountain-building processes far more active than today. The same must have been true on other bodies of the solar system, including those from which the meteorites have come.

Despite persistence of the view that the meteorites, asteroids and perhaps comets are the debris of a former planet, it has not been universally accepted for several reasons, a prime one being the dearth of material in the asteroid belt. Some 1,660 asteroids have been catalogued, the largest being Ceres, with a diameter of about 480 miles. Probably 55,000 come within observable range at one time or another. The larger ones reflect sunlight with constant intensity, implying that they are spherical and hence formed in the same manner as the planets and moons, whereas some of the smaller asteroids vary in brightness as they are seen from different directions, suggesting that they are angular fragments of a larger body or bodies. However, even if all the asteroids were lumped together, they would constitute only a few hundredths of the mass of the moon. Taking the rate of meteorite falls on the earth as a clue to the rate at which material has been escaping from the asteroid belt throughout the history of the solar system, there would not have been enough material there at the start to build a planet of even moderate size.

Another question is: How could a planet blow up into fragments? One suggestion has been that two planets collided, but it has been calculated that the energy released, as heat, by their falling together would be more than enough to vaporize both bodies. Only a few pieces would be thrown free to tell the tale. It has also been proposed that an earlier civilization blew up itself and its planet. People fearful of nuclear war talk of blowing up our own world, not realizing that to do so would require a great deal more than dropping a hydrogen bomb down a volcano. The necessary force would dwarf the wildest dreams of weapons designers. The explosion would probably have to take place in the heart of the earth and would have to be powerful enough to push the chunks so far apart that they would not fall back together again. As with a collision, an explosion of this magnitude presumably would vaporize the planet. Furthermore, the dating of meteorites indicates that such a suicidal civilization would have had to evolve far faster than our own.

It has been the discovery of new dating techniques that, in the past few years, has made it seem possible to reconstruct the history of the meteorites and thereby seek clues to the sequence of events that gave birth to the solar system. By determining the extent to which various radioactive isotopes within a meteorite have decayed into other substances, a suprisingly broad range of dates in the specimen's history can be estimated (with varying degrees of reliability). These include:

1. The time when the sample solidified from the molten state.
2. The time when it cooled to temperatures such as those in the earth's crustal rocks.
3. The time when it was ejected from its parent body.
4. The time of its fall to earth.
5. Its size and shape before erosion and fragmentation by its flaming plunge through the atmosphere.

The isotopes that figure in all of these calculations are varieties of each element that differ only in the number of neutrons in their nuclei. For example, xenon gas, a minor component of our air, exists in nine forms, or isotopes, with 'mass numbers' ranging from 124 to 136. The mass number represents the total number of protons and neutrons in the nucleus. The number of protons for any given element, such as xenon, must always be the same, since it determines the electric charge of the nucleus and hence the chemical properties of the atom. It is therefore the quota of neutrons that varies from one isotope to another.

Apart from hydrogen, the elements, including all their isotopes, are thought to have been formed by nuclear reactions within stars or in cataclysmic stellar explosions, such as the supernovae. Since all atoms of the solar system were presumably brewed in the same stellar 'pots', the abundances of these isotopes, thoughout the system, are thought to have been the same originally. Thus, for the solar system and possibly on a far larger scale, there seems to be a characteristic distribution of isotopes. All else being equal, the proportions of various isotopes in a sample of xenon from Jupiter should be the same as in a sample of that gas on the earth.

Any deviation from these characteristic abundances means that some disturbing influence has been at work. One such influence is radioactive decay, whereby atoms emit radiation and transform themselves into a lighter element. Some radioactive isotopes tend to decay rapidly. A few do so, on the average, very slowly. The moment when any one atom in a specimen of radioactive material will decay cannot be predicted, but the over-all rate, for billions of atoms, can be determined by experiment. This rate is expressed in terms of 'half life', which is the time it takes for half of a given amount to decay.

It has been found that various decay processes in the meteorites are, in effect, stop-watches set in motion by events of the distant past. Some, for example, can be used to determine when the material passed from the molten to the solid state, since the decay products prior to that would have dissipated. Among the most useful of these decay 'modes' are those of rubidium 87 into strontium 87, of rhenium 187 into osmium 187 and of uranium 238 into lead 206. So slow is the rate of these decays that they can be used to reckon time in billions of years.

The analysis of many meteorites for all three of these decay modes points to an end of their molten phase about 4·5 billion years ago. This is roughly one billion years before formation of the oldest rocks found, so far, on the surface of the earth. Hence the meteorites are of enormous interest to those seeking to learn what happened in the earliest stages of the solar system's formation. At first glance this evidence that the meteorites were molten when the solar system was young might seem incompatible with the idea that the system originated from cold dust and gas, but it is suspected that, early in its history, the sun burned more brightly than today and also, as mentioned earlier, that the bodies of the solar system were much hotter, internally, from heating by radioactive elements.

The cooling-off schedule of the meteorites has been reconstructed by examining decay processes that produce gases such as helium, argon, and xenon. As long as the material was hot, these gases diffused through it and vanished. However, as it cooled, the gases were trapped and retained for the edification of future scientists.

The most widely used of these decay modes is that of potassium 40 into argon 40. Apparently the specimens began accumulating the argon produced by this reaction once they had cooled below a few hundred degrees. This would be when the body from which the meteorite came first cooled in the childhood of the solar system. However, if the event that subsequently tore the fragment loose was sufficiently violent, the specimen would be reheated and the argon would escape; whereupon its accumulation would again start from zero.

In 1960 John H. Reynolds at the Berkeley campus of the University of California found that some meteorites contain an abnormally large amount of xenon 129, in proportion to other xenon isotopes. It was, in fact, 50 per cent more plentiful than in the xenon of our atmosphere. He concluded that the excess xenon 129 was a decay product of iodine 129, a radioactive substance that, because its half life is only 17 million years, presumably disappeared when the solar system was young.

The beauty of xenon 129 is that it makes it possible to estimate how long a meteorite was incorporated within some large body, such as an asteroid or planet. This is because it would begin to accumulate in the sample as soon as the latter cooled below 200 degrees. Take for example the stony meteorite that showered fragments on Bruderheim, Alberta, in 1960. It apparently began storing xenon 129 some 4·5 billion years ago. This, on the basis of other radio-dating analyses, was less than 100 million years after it solidified.

The length of time that a meteorite has flown through space in lonely splendour after being torn from the heart of its parent can be determined by measuring the effects produced within it by cosmic rays. The latter comprise the most penetrating form of radiation known. They are, primarily, not rays of light, but extremely high energy particles, raining on the solar system from all directions. Some even penetrate deep enough into the earth to be detectable in mines, but, fortunately for us all, by far the greater proportion of them are diverted by the earth's magnetic field into nearby space, or are halted by the blanket of air over our heads.

These rays transmute the atoms that they strike in ways that can be predicted, and it is this that makes it possible to use the exposed material as a timekeeper. By measuring the extent to which cosmic rays have generated stable substances (those that are not radioactive), it is possible to estimate how long the object was exposed to unmitigated bombardment. This period began when the object was ejected from within its parent body and ended when it fell under the protective umbrella of the earth's air and magnetism.

Likewise, the date of the fall to earth can be estimated. Since cosmic rays generate both stable and radioactive substances, it is possible, from the amount of stable material found in a specimen, to estimate how much of the unstable (radioactive) stuff was there when it fell. The extent to which the latter has decayed is an indication of how long ago the fall occurred.

In this way it has been calculated that a 2,900-pound meteorite found in Mexico in 1872 fell there about 800,000 years earlier. The oldest known fall, as determined by this method, is that discovered at Ider, Alabama, in 1961. It had apparently lain there some three million years.

In April 1964 some of the latest attempts to read meteorite stopwatches were reported at a symposium on the subject at the National Academy of Sciences in Washington. Edward Anders described recent analyses in the United States and Germany of cosmic-ray effects on meteorites. They showed that in a large portion of cases the ages of the specimens as small bodies fall into specific categories. Meteorites of similar composition seem to have come from the same parent bodies in that their break-up dates tend to coincide. The typical break-up dates that he gave, in years before the present, were:

Iron	600,000,000
	900,000,000
Aubrites	45,000,000
Bronzite chondrites	4,000,000
Hypersthene chondrites	25,000
	3,000,000
	7–13,000,000
	16–31,000,000

The last of these categories, the hypersthene chondrites, constitute more than half of the meteorites that have been seen to fall. What struck Anders was that many of these meteorites appear to have been storing gaseous products of radioactivity (helium 4 and argon 40) for only 400 million years. Other meteorites characteristically cooled enough to retain these gases soon after formation of the solar system, some 4·5 billion years ago. The explanation proposed by Anders was that the hypersthene chondrites, like the other types, cooled when the solar system was young, and began storing helium and argon gas. But, some 400 million years ago, the asteroid of which they formed a part collided with something. The resulting heat liberated the helium and argon stored up to that time, and the stop-watch, so to speak, started over again from zero. Following this catastrophic collision the resulting fragments banged into one another from time to time, not generating enough heat to interrupt the accumulation of gases but breaking the chunks into smaller pieces. This would account for the diversity of ages indicated by the cosmic-ray effects, since the latter are not significant more than about two feet below the surface.

Anders, looking around for possible remnants of the parent body, pointed out that chunks of material circling the sun within the asteroid belt are unlikely to suffer changes of orbit sufficiently radical to bring them in near the earth. Collisions along the 'dangerous freeway' of the belt must be frequent, but they would tend to spread the material out along the orbital path rather than produce major changes in that path.

However, it had been pointed out by James R. Arnold of the University of California at La Jolla that if a chunk from the asteroid belt is in an orbit that swings in close enough to cross the orbit of Mars, then it can be seized by the gravity of that planet and thrown into a new orbit penetrating the inner solar system as far as the earth. Of the 1,660 catalogued asteroids, 34 cross the orbit of Mars and, Anders noted, these 34 tend to fall into four families, each with its own orbit. Three of the families bear names: Phocaea, Desiderata and Aethra. In each case the members of the family are presumably fragments of a single parent body, there being countless additional fragments that we cannot see. The largest of these parents, from the pieces that are visible, must have been at least 140 miles in diameter. Anders proposed that one of these bodies broke up 400 million years ago and is the source of more than half the meteorite falls.

There are also eight 'Apollo' asteroids in orbits so elongated that they reach into the solar system as far as the earth. They are small, ranging in diameter from less than a mile to Eros, a seemingly irregular chunk 20 miles long and from 5 to 10 miles wide. Its next close approach to the earth is due in 1975. These objects are in orbits similar to that of the only well-documented meteorite – that which fell on Czechoslovakia in 1959. The Czechs were carrying out studies of meteor trails, using cameras spaced 25 miles apart. A fast-spinning drum in one camera closed the shutter for a brief fraction of a second on each rotation. This chopped the streak left on the photograph by a meteor into segments that showed its speed. The cameras also made it possible to calculate the flight path, using distant stars to establish three-dimensional geometry. On the night of 7 April the sky blazed with a fireball which, by good fortune, was recorded by both cameras, enabling Zd. Ceplecha of the Czechoslovak Academy of Sciences, for the first time in history, to reconstruct the orbit of a meteorite with precision. On the basis of his calculations, the

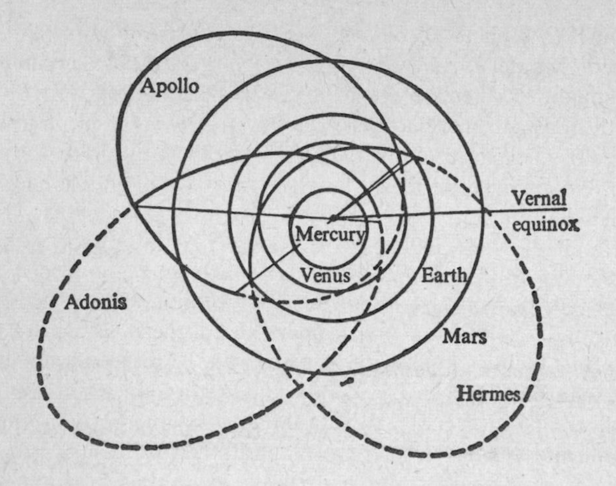

The orbits of three Apollo asteroids reach into the centre of the solar system well past the orbit of the earth. The portion of each path lying below the plane of the earth's orbit is shown as a dashed line. (From F. G. Watson, *Between the Planets*, rev. ed., Harvard University Press, 1956.)

object itself was found in a field near the town of Příbram, and he showed that its orbit reached out to the asteroid belt, between the orbits of Mars and Jupiter.

At the 1964 symposium Anders proposed that objects such as the Příbram meteorite and the Apollo asteroids may be fragments thrown towards the earth by the gravity of Mars – fragments from one of the 'four families' that swing in close to the orbit of that planet.

Arnold, who was at the symposium, challenged this analysis. He suggested that some of the eight objects that come close to the earth may be old comet heads and cited the fearful 1908 explosion of the Tunguska 'meteorite' as the probable impact of a comet. Let us hope, he said, that the next such impact occurs during a period of relaxed international relations, for it would closely resemble the explosion of an outsized hydrogen bomb. He pointed out that the movements of bodies that periodically cross the orbital highways of the solar system are complex and unpredictable, since they always face the possibility of passing close enough to a planet to be swung into a new path. Using the so-called 'Monte Carlo' method of analysis on a large computer, he sought to calculate the odds that a meteorite, arriving at the earth in a path like that recorded by the Czechs, had originally come from the asteroid belt. He found that, taking into account possible planet-encounters throughout its orbital lifetime, its origin could just as well have been the moon.

He cited a proposal by Harold Urey that most of the stony meteorites are chips knocked off the moon when that body is hit by large iron meteorites. On the basis of the computer analysis, he said, Urey's hypothesis 'must be taken seriously'. Urey had pointed out there is a gross difference between the typical ages of irons as small bodies (600 to 900 million years) and those of the stones (as little as 25,000 years in the case of the Farmington meteorite). His view, of several years' standing, was that the irons that occasionally hit the moon and earth come from the asteroid belt and are veteran space travellers, whereas the stones reside in space, subject to cosmic rays, only while they are playthings of the vital gravities of earth and moon. Once earth's gravity wins out, they plunge into the atmosphere like a space capsule on re-entry.

The composition of the meteorites clearly has a tale to tell, full of sound and fury, but there is still little agreement on what it signifies. When specimens are sliced and polished, they show structures of great variety and beauty. Sometimes a meteorite of one type will be completely embedded within a meteorite of quite a different type. There is evidence, in some cases, that the material within one specimen represents five 'generations' of break-up and reconsolidation.

The most perplexing of all features in the stone meteorites are the strange chondrules. So uniformly are they distributed through some specimens that when polished in cross section the material resembles a well-mixed rice pudding. Such is the challenge of the chondrules that there seems no limit on the imaginations of those seeking to explain them. It has been proposed that they are droplets of once-molten rock ejected during volcanic eruptions of a sort unknown on earth (where chondrules are not associated with volcanoes). Some have said that they were produced within the rock by the shock of collision. Or, it has been suggested, they are frozen drops of material that was vaporized by an explosive collision, and then condensed into a fiery rain. Another proposal is that they may have condensed from gases spun off by the sun in its formative, fast-rotating stage. Indeed, part of the intense interest in meteorites lies in the clues that they may hold to such birth agonies of the solar system.

Urey believes that most, if not all, features of the stony meteorites can be explained by the effects of bombardment and other processes on the surface of the moon. If he is correct, his hypothesis will be dramatically confirmed when astronauts first sample rocks of the lunar surface.

Directly pertinent to the question of whether or not the meteorites carry evidence of life is the size of their parent body or bodies. Here the views are almost as divergent as on the question of chondrules. Urey has continued to maintain that they must have been at least as large as the moon, with the added weight of a large, primordial envelope of gas, to crush carbon into diamond crystals. At the 1964 symposium he argued that the larger bits of diamond found within meteorites could not have been formed by shock. He was supported, to some extent, by Clifford Frondel, Professor of Mineralogy at Harvard, who suggested that the cubic crystals of graphite found within meteorites may once have been pressure-formed diamonds that were later altered by some process.

Anders, on the other hand, argued that the typical parent bodies must have been less than 125 miles in radius. Otherwise, he said, some of the specimens could not have retained the gases produced by their radioactivity throughout the past four billion years. The fact that they have retained these gases since the solar system was young means their parents cooled off rapidly. While they were hot, as noted earlier, the gases could not accumulate. If the bodies cooled early, Anders argued, they must have been much smaller than the moon.

There are those, too, who are loyal to the idea that the meteorites are daughters, or granddaughters, of one or more planets. Notable among these are J. F. Lovering and Albert E. Ringwood at Australian National University. In fact, the only point of agreement among the various speculators is that no simple theory can explain all the features of these objects. As Arnold put it at the 1964 symposium, 'There are serious objections to all present models.' And yet, he added, 'the meteorites must come from somewhere'.

Our chief hope for determining whence they come is a newly established network of sixteen stations in the central plains of the United States. With sixty-four cameras constantly monitoring the night sky, they should record an average of at least one fireball a year, making it possible to find such meteorites on the ground with greater ease, as well as to reconstruct their orbits. Special attention will be focused on those specimens that are found to have been exposed to cosmic rays for comparatively brief periods – less than a million years. As Arnold pointed out at the symposium, the chances would be good that such an object had not yet been yanked out of its original orbit. Thus its path, if reconstructed, should pass through its point of origin.

Meanwhile the absence of a generally accepted theory for the meteorites makes it particularly difficult to assess the claim that life existed on the body or bodies on which some of them originated. Current efforts at such an assessment are the subject of the next chapter.

11

'*Wax and Wigglers*'

BEGINNING in 1959, the investigation of meteorites took a sensational turn. In that year Melvin Calvin, and those of his co-workers interested in the chemical evolution of life, obtained from Edward P. Henderson at the Smithsonian Institution in Washington a chunk of the carbonaceous chondrite that had fallen near Murray, Kentucky, in 1950. As they had done with the material produced in their experiments with 'primordial' gases, they subjected the meteorite to the most precise techniques of analysis at their disposal and, in January 1960, Calvin and Susan K. Vaughn reported their results to the First International Space Science Symposium, held in a municipal hall facing the Mediterranean at Nice.

The title of their paper was: 'Extraterrestrial Life: Some Organic Constituents of Meteorites and Their Significance for Possible Extraterrestrial Biological Evolution.' As they themselves pointed out, it was remarkable that the first symposium held to exchange the initial fruits of rocket and satellite exploration should include such an item.

'The mere fact that the subject of extraterrestrial life was deemed suitable to be included in this meeting,' the two scientists told the space researchers from East and West, '... is evidence enough that the interest of men in the possible living population of those bodies they see above them is still alive, as it has been ever since men first looked and thought about those bodies.'

They cited the yearning of scientists for a bit of the moon or some other celestial body to see if it harboured the pre-life chemicals that presumably existed on earth before life emerged and devoured them. If these could be found, it might indicate the chemical paths whereby life did, in fact, evolve. Calvin and Miss Vaughn then pointed to the steady rain of meteorites and said: 'Here we have a gift from heaven, so to speak, something which we do not have to go out and get ...'

They noted that the first contemporary attempt at analysis of the carbonaceous chondrites was that of G. Mueller at University College, London. In 1951–2 Mueller found material in Cold Bokkeveld resembling hydrocarbons – the family of compounds typical, for example, of petroleum products. Actually, this had been done, in a cruder way, as early as 1868, when Pierre Eugène Marcelin Berthelot, the great French chemist, analyzed a sample of the Orgueil meteorite that had fallen four years earlier. He obtained hydrocarbons, both liquid and gaseous, so like those of petroleum that he described them as 'a new analogy between the carbonaceous materials of the meteorites and the carbonaceous materials of organic origin' on earth.

Calvin and Vaughn described Mueller's results as 'a very exciting business'. It is clear, they continued, that 'there are, floating around in outer space, already rather complex carbon compounds.' In other words, it was not necessary to depend entirely on processes that took place on the primitive earth to obtain the complex chemicals from which life arose. Some were formed in space and arrived on the earth ready-made.

Calvin's and Vaughn's paper explained how samples of the meteorite had been vaporized and run through a mass spectrometer designed to show the relative masses of molecules that pass through it. A beam of electrons is fired at the sample, ionizing, or imparting an electric charge to, its molecules. These are then shot through a magnetic field that bends their flight path. Because the extent to which the path is bent is determined by the weight of each molecule, the stream can be analysed for its abundances of molecules of different weights.

Five solvents, including water and carbon tetrachloride, were used to extract materials of various kinds which were then analysed in terms of their absorption of infra-red and ultra-violet light. The carbon tetrachloride extract showed the infra-red spectra of methyls and methylenes, and the ratio between them suggested the presence of molecules built on chains of fifteen or more carbon atoms. The water extracts were studied under ultraviolet light and the variations of their spectra, at different acidities, were those typical of cytosine, one of the four bases that carries the 'code of life' in the DNA molecule.

They did not claim that the substance was, in fact, cytosine, but rather that it resembled that chemical closely. (Some have recently suggested that what may actually have been detected was cytosine derived from resin used in the separation process.)

Of special interest, the two scientists said, was the fact that the variety of complex substances, like cytosine, in the meteorite seemed to be limited. In other words, chemical evolution elsewhere, as postulated by Calvin for the earth, appears to have been selective, instead of producing a hodgepodge of chemical combinations. However, a rather surprising discovery was the seeming absence of those other key, pre-life compounds, the amino acids.

The California scientists also found a mixture of hydrocarbons that appeared similar to those in waxes and petroleum, although they could not specifically identify any one of them. They assumed that all these substances had been formed by non-living processes, but to them, as to all those concerned with the origin of life, this evidence that such processes had actually taken place elsewhere was an exhilarating vindication of their speculations.

It was in the following year that the debate over carbonaceous chondrites moved into high gear. On 16 March 1961 a paper was presented to a joint meeting of the chemical and geological sections of the New York Academy of Sciences describing a new analysis of the Orgueil meteorite that had fallen on France at the height of Pasteur's battle against spontaneous generation. The authors were Bartholomew Nagy and Douglas J. Hennessy, both of the Department of Chemistry at Fordham University's Graduate School of Arts and Sciences in the Bronx, and Warren G. Meinschein, a petroleum chemist at the Esso Research and Engineering Company in Linden, New Jersey. Nagy, aged thirty-four, had received his early education in his native Hungary and more recently had done research for an American oil company.

They described the results obtained when a bit of the meteorite was subjected to techniques typically used in petroleum research. Hydrocarbons were extracted from the sample by distilling and were passed through a mass spectrometer at the Esso laboratories in Linden. Some of the hydrocarbons proved to be built on chains of as many as twenty-nine carbon atoms and displayed striking similarities to the paraffins and other hydrocarbons found in living matter. Such biological hydrocarbons tend to have carbon spines of odd-numbered atoms (21-atom chains, 23-atom chains, etc.). The experimenters found that the assortment of paraffins in the Orgueil meteorite, when classed by weight, resembled that in butter and in sediments containing recent life remains. One constituent seemed related to the cholesterol found in blood, they said.

They concluded that they were not analysing bits of dirt that had entered the sample after its fall to earth, for the specimen's saturated hydrocarbons were far more abundant than in typical soils. Saturated molecules are those whose carbon atoms lack free sites where additional hydrogen atoms could hook on.

The three scientists described their efforts to avoid contamination: glass equipment acid-cleaned and baked in a vacuum; mortars and pestles kept red-hot for at least a half hour; no grease used on stopcocks, and so forth.

The next day Robert K. Plumb, reporting in the *New York Times*, quoted one of them as saying: 'We believe that wherever this meteorite originated something lived.' It was clear that they regarded this as the first direct evidence of life beyond the earth. Plumb got in touch with two men experienced in the study of meteorites: Harold Urey in California and Brian Mason at the American Museum of Natural History in New York. Both urged caution in the interpretation of this evidence. Nevertheless, the fats – or rather, the saturated hydrocarbons – were in the fire, and the conflagration was not to simmer down for a number of years.

With the surge of public interest that followed the New York meeting a reporter from Science Service went to see Henderson, keeper of the meteorites at the Smithsonian, and discovered that, in addition to giving Calvin a chunk of the Murray fall, he had presented another piece to a microbiologist at the United States Geological Survey named Frederick D. Sisler. The latter was cultivating living organisms that he had ostensibly extracted from the heart of the specimen.

Needless to say Sisler's phone began ringing and the press quoted him as saying that, while he could not guarantee the extraterrestrial origin of his organisms, this was a real possibility. He pointed out to the reporters, however, that water seepage could have carried bacteria into the heart of the porous specimen. *Life* magazine published an article entitled, 'Wax and Wigglers: Life in Space?' The 'wigglers' were Sisler's bacteria; the wax was the paraffin found by Nagy and his colleagues.

Sisler had done his research in conjunction with Walter Newton, chief of the Germ-Free Laboratory of the National Institutes of Health, the complex of federally operated research centres at Bethesda, Maryland. Sisler, who had previously been on the Institutes' staff, was now head of the Geological Survey's Special Studies Unit.

A few weeks after news of his findings broke in the press, he was invited to describe them at a Lunal and Planetary Exploration Colloquium, organized by North American Aviation, Inc., at its headquarters in Downey, California. Harold Urey was on the steering committee and those in attendance included most of the figures soon to be caught up in the controversy as to whether or not there is evidence of life in the carbonaceous chondrites.

For about five years, Sisler said, he had been studying meteorites, concentrating on specimens provided by the Smithsonian Institution. 'I have found either traces of organic matter or germs, or what appeared to be germs living or dead, in almost everything I have examined,' he said.

His procedure with the two Murray specimens was as follows: The surface was sterilized by exposure to intense ultra-violet radiation, from all sides, for half a day. It was soaked in hydrogen peroxide to rid it of loose dirt. It was then held briefly over a flame and was plunged into a germicidal solution. Finally it was placed in a germ-free chamber.

The chambers are tank-like containers within which everything, including the air, is sterile. They are inhabited by mice that are germ-free even to the extent of having no intestinal flora, such as the bacteria that live harmlessly in our digestive systems. These mice serve, in part, as monitors. If there is any contamination of the chamber, this will immediately show up in their intestines. Rubber gloves sealed to the walls of each chamber make it possible for researchers to work inside it without coming into direct contact with the sterile contents.

Sisler ground up the samples with a mortar and pestle and inoculated some of the mice with material obtained from the interior of the specimen. The mice were unaffected, but when such material was placed in a clear fluid rich in nutrients, the latter sometimes became cloudy. It often took several months to do so, which suggested to Sisler that the sparks of life in the meteorite might have been crippled in some way, requiring a long time to reconstitute themselves. The bacteria looked, in the microscope, like segments of a corkscrew. Sisler could not recognize in them any common form of terrestrial bacteria but, unlike some of those reporting on his work, he resisted the temptation to say they were 'like nothing on earth'.

He found that the meteorite bacteria were able to grow both in the absence of oxygen and its presence. That is, they were 'facultative anaerobes', like some forms of terrestrial bacteria. At the California colloquium he listed four possible explanations for his discovery of living bacteria: his analytical method may have been faulty; earth germs may have penetrated the specimen; there may have been life on the body whence this fragment came; or there may have been 'autocatalysis' in which the meteorite material, immersed in a favourable growth medium, completed the final stage of chemical evolution and came to life – 'something akin to spontaneous generation'.

Sisler was not the first to report finding bacteria in meteorites. In 1932 Charles B. Lipman, in a publication of the American Museum of Natural History, described a series of experiments that resembled those of Sisler in that he took what he considered drastic measures to sterilize the exterior of each specimen and ground it up under what he felt were sterile circumstances, although he did not have at his disposal the elaborate facilities of the Germ-Free Laboratory in Bethesda.

Lipman was no fly-by-night crackpot. Having come to the United States from Russia as a child, he had risen to be Professor of Plant Physiology and Dean of the Graduate Division at the University of California in Berkeley. From pulverized fragments of several stone meteorites (Pultusk, Johnstown, Modoc and Holbrook) he cultured a variety of bacteria, including cocci, chains of cocci, rods, chains of rods and chains of sausagelike bacteria. He conceded that they resembled earth bacteria, but said there was no valid reason why bacteria like our own could not have evolved elsewhere. He anticipated the argument that any bacteria from space would have been killed by the heat of the meteorite's flaming passage through the atmosphere. Actually, he said, the heat thus generated has time to penetrate only a fraction of an inch below the surface, an observation now well substantiated. For example, when a stone fell at Colby, Wisconsin, on a hot July afternoon in 1917, frost was forming on its surface by the time local residents reached it a few minutes later. The frigidity of space, within the heart of the stone, had already banished its surface heat.

Lipman believed that he had effectively ruled out any contamination. The evidence convinced him, he said, that stone meteorites 'bring down with them from somewhere in space a few surviving bacteria, probably in spore form but not necessarily so . . .'

Most scientists were sceptical, and in some cases indignant. Michael A. Farrell of the Department of Bacteriology at Yale University, writing in a subsequent issue of the museum's organ, said the newspaper publicity following Lipman's announcement 'cannot but be disturbing to the minds of earnest searchers for truth, especially when the supposed findings fail of corroboration in other laboratories.' He cited Lipman's earlier report of reviving bacteria that had allegedly lain dormant for millions of years in anthracite coal. Others, Farrell said, had been unable to duplicate Lipman's results except in coal whose cracks furnished routes whereby contaminating ground water could seep in.

Until more convincing evidence is brought forth, said Farrell, 'Lipman's excursions into the field of life beyond this globe must be considered as a flight of imagination through space.'

Sisler, more cautious than Lipman, readily conceded the possibility of contamination and, in fact, told the author that he believed this to be the most probable explanation. At that time (1963) he was still culturing his meteorite specimens in the belief that some might come to life only after several years. Also, he said, a large number of experiments were needed to show, statistically, whether or not the organisms were, indeed, contaminants. He stressed, however, that the possibility of extraterrestrial origin has not been eliminated.

The publicity given Sisler's work sharpened the debate on the observations of Nagy and his colleagues in New York. Many sceptical voices were raised, but it was Edward Anders of the University of Chicago who emerged as spokesman of the doubters. His debate with the New York group began at a meeting of the American Geophysical Union in Washington soon after the session at the New York Academy of Sciences, and he again confronted Nagy's associate, Warren Meinschein, a few days later at the meeting in Downey, California, where Sisler told of his live cultures.

As evidence that saturated hydrocarbons such as those found in the Orgueil meteorite were unlikely to have originated except through life processes, Meinschein reviewed the efforts to explain the origin of petroleum. He pointed to the nineteenth-century arguments that 'mineral oils' could have somehow been formed by the action of heat and pressure on inorganic substances buried deep in the earth, noting that no reasonable way had been found in which this could have come about. The conventional view today is that crude oil has been derived in some way from sediments rich in the debris of aquatic life.

Meinschein told how, in the 1940s, the American Petroleum Institute launched 'Project 43', in which six major universities worked on the problem for ten years without finding a satisfactory explanation for the creation of petroleum. It was found, for example, that if the fatty substances in fish are distilled, a combination of hydrocarbons much like that in petroleum is produced. But it seems evident that the oil-bearing sediments have always been cool. How could the fish remains have been distilled? Likewise, it was suggested that bacterial action or radiation from the surrounding rocks could have transformed the debris of prehistoric life into the composition of petroleum, but those theories did not stand up either. So desperate were scientists for an explanation that they even gave consideration to a proposal that all of our petroleum products, from gasoline to tar, have fallen from the sky. This was advanced in 1960 by Frits W. Went, a Dutch-born botanist who was Director of the

Missouri Botanical Garden and a member of the National Academy of Sciences. He proposed that the blue haze of resinous fumes given off by plants, particularly conifers and sagebrush, rises high enough to be altered by ultra-violet sunlight and is then brought back to earth by rain, accumulating in the sediments near river mouths. The publication of this unorthodox theory in the *Proceedings* of the academy was a reminder that we still do not know the source of the fuels that drive our planes, ships, trains and cars.

Meinschein's argument was that study of the problem, such as that in Project 43, while unable to advance a solution, had shown that the origin of petroleum must have been biologic. Hence, he said, the hydrocarbons in meteorites must similarly be the products of life. (However, dissent continues to be published in Britain, the Soviet Union and elsewhere supporting non-biological production of petroleum hydrocarbons.)

Among the questions raised by Anders was whether the similarity of the Orgueil hydrocarbons to those in fuel oil might not be more than a coincidence. He noted that the meteorite is quite porous and hence 'breathes' whenever there is a change in barometric pressure. The specimen had lain for almost a century in the American Museum of Natural History in New York and, Anders wrote in an analysis, 'There are not many places on earth that burn and vaporize fossil fuels [coal, oil, gasoline, etc.] at a higher rate than New York City.'

In other words, he said the meteorite had been breathing exhaust fumes since automobiles first began riding the streets of New York and some of the material must have stuck to the inner pores of the specimen. Only a very small amount would be needed, Anders said, to produce the characteristic spectrum of hydrocarbons. Meinschein and his colleagues countered by saying that they had found the meteorite 100 times richer in hydrocarbons than could be explained by any such contamination.

What has prejudiced scientists against the arguments of Nagy, Meinschein and their colleagues throughout the debate has been the seeming unlikelihood that life could originate on a body in the asteroid belt. In discussing the report of the Fordham group, J. D. Bernal, who had pondered long on the problems of life's origin, wrote in the *Times Science Review* that the discovery of hydrocarbons in the Orgueil meteorite might be taken to imply that at some time between 2 and 200 million years ago there was a planet capable of sustaining life in some form, at least in the chemical sense. The trouble, he said, was to find a suitable home for this planet in the solar system. If it were inside the earth's orbit, the meteorite material would have been altered by the sun's heat. Such an alteration had clearly not taken place. If it were in the belt of asteroids beyond the orbit of Mars, it would be extremely cold. The asteroid belt, Bernal pointed out, receives only about one-tenth as much solar heat as does the earth. It is therefore hard to see, he said, how life could survive in such an environment, much less go through the slow, delicate process of originating. Hence he proposed that, unless the carbonaceous chondrites come from another solar system, the most obvious source of the hydrocarbons was the earth itself.

Thus Bernal suggested the possibility of contamination. The idea that the Orgueil hydrocarbons had been produced by life in space was the reverse of a concept set forth earlier by Bernal, that the compounds native to the meteorites were formed without intervention by living forms and helped seed the primitive earth with material that figures in the evolution of life. He cited the proposal of Brian Mason that the carbonaceous chondrites are relics from the earliest stage of the solar system, their complex compounds having been synthesized by such stimuli as ultraviolet light and high-energy radiation from the youthful sun. Calvin too, as reported earlier, had concluded that hydrocarbons in the Murray meteorite were formed in such a manner.

When word of the Orgueil discoveries reached New Zealand, Michael H. Briggs of Victoria University in Wellington submitted a sample of the only such meteorite that has fallen on New Zealand, the Mokoia, to an elaborate series of analyses. He reported evidence of purines in the material, but felt he could not specify which of the purines were there. Two purines, adenine and guanine, help carry the code of life. Briggs felt his inability to detect amino acids – ubiquitous substances in earth life – ruled out any serious contamination of the specimen.

He concluded that the compounds in Mokoia were either

decomposition products of extinct life or had formed naturally. He cited the proposal of Joan Oro (described in Chapter 9) that comet heads striking the earth may have enriched the chemical diversity of our planet's surface in its early days. Comet orbits traverse the asteroid belt, which is so cluttered with material that collisions there should be frequent, thus enriching the asteroids in the same manner.

At the Downey colloquium Anders, too, argued that the hydrocarbons in Orgueil might have been formed in space by non-biological processes. He concluded his discussion with a dig at the publicity accorded the Orgueil findings: The only connexion between meteorites and life, he said, 'is that an article on meteorites appeared in a magazine called *Life*.'

While the announcement of Nagy and his colleagues that they had found biological substances produced a stir, their report of a few months later created a sensation. In the issue of *Nature* for 18 November 1961 Nagy and George Claus of New York University Medical Center said that they had examined samples of four carbonaceous chondrites and in all of them had found microscopic particles resembling (but not identical to) fossil algae of the kind that live in water. Two other stony meteorites, examined for comparison, showed no such 'organized elements', they said. They described five types of organized elements – that is, fossil life forms – that appeared in the carbonaceous samples. Some of them seemed to have perished in the midst of 'cell-divisions', said Nagy and Claus, although they themselves put the expression in quotation marks. The 'Type One' objects were small and circular with double walls. The material inside those walls stained diffusely in a manner typical of certain cells. The 'Type Two' objects were like those of the first category except some of them were covered with spines or showed other appendages. 'Type Three' objects were shaped like shields. 'Type Four' objects were cylindrical and 'Type Five' objects appeared to be six-sided with tubular protrusions on three sides. Only one of these was seen, initially. It was the most strikingly lifelike of the five types.

All told, the forms were plentiful in the Orgueil and Ivuna meteorites, but were not so well-defined in the Murray and

Mighei samples. They did not resemble any known mineral particles, the two men said. While they also found some obvious contaminants – well-known forms of terrestrial bacteria and algae – they felt their 'fossils' must be indigenous to the meteorites. Except for the single specimen of Type Five, all the others resembled Dinoflagellates or Chrysomonads. The former, even though they are plants, propel themselves through the water. None of these forms live in the soil and hence they were viewed by Nagy and Claus as unlikely contaminants. Furthermore, whereas Orgueil fell in the south of France, Ivuna fell in an arid, tropical region of central Africa. It seemed unlikely that similar contaminants would enter both. They said that they had interpreted the organized elements 'as possible remnants of organisms'.

Coming on top of all the other discoveries, by Calvin, Briggs, Meinschein, etc., this report impelled almost everyone with access to a carbonaceous chondrite to take a close look at it. Among these was Robert Ross at the British Museum in London. The museum possessed some fragments of Orgueil, and a number of the Type One forms described by Nagy and Claus were found in them, according to Ross. Likewise, he reported two with a mushroom shape, on a microscopic scale, and objects resembling collapsed spore membranes. He and his colleagues felt there was a strong indication that these were evidence of life somewhere in space, but they conceded that there was, as yet, no proof.

In Calgary, Alberta, Frank L. Staplin of Imperial Oil Ltd subjected a sample of Orgueil to the kind of pollen analysis used by oil men to assess samples of sedimentary rock drilled from deep in the earth. The fossil pollen grains are a clue both to the age of the rock and the climate in which it was laid down. Staplin found a number of objects that, in size, texture and acid resistance, 'superficially' resembled some of the one-celled algae. Because of their importance in oil prospecting, he said, the pollen grains in deposits less than 600 million years old are well known. In general the Orgueil forms were clearly different, and he identified two entirely new genera, or general categories of plant, to which, according to custom, he appended his own name: *Caelestites staplin* and *Clausisphaera staplin.* The first name was clearly intended to denote its heavenly origin.

Another he identified as *Protoleiosphaeridium timofeev*, a fossil pollen grain discovered in terrestrial rocks by Boris Vasilyevich Timofeyev, pollen specialist at a Soviet government institute in Leningrad that does research on oil prospecting.

Meanwhile, Timofeyev was himself looking at a carbonaceous chondrite - one that fell at Mighei, near Odessa, in 1889. He centrifuged a sample of its material and in the lighter fraction found a number of rounded objects that he felt resembled the oldest form of alga known, the Protosphaeridae, although they did not actually fall into any earthly classification.

Like Staplin, he hung his name on various of these presumed relics of life elsewhere and when he heard of Staplin's work on Orgueil, he named one of the Mighei 'fossils' *Prototrachysphaeridium staplini timofeyev*. In May 1962 an All-Union Conference of Astrogeologists was held in Leningrad and Timofeyev told of his findings. A number of the Soviet scientists present were sceptical, but Timofeyev persisted in his view that the grains were evidence of life having existed elsewhere in the universe.

When a conference on pollen studies was held in Arizona that year, Timofeyev was unable to make arrangements to go but an abstract of the paper that he intended to present was published by the conference organizers. When I visited Leningrad, late in 1962, Timofeyev took me to his apartment to squint at microscopic preparations of Mighei material. The Soviet scientist bristled with enthusiasm, clearly convinced of the validity of his finds.

Confirmations of the Nagy-Claus report seemed to be coming in from right and left. P. Palik at Eötvös Lorand University in Budapest, Hungary, asked for a bit of Orgueil and found in it six filament-like structures reminiscent of algae. She felt some of them 'may possibly be indigenous to the meteorite'.

Nature, which traditionally opens its pages to controversy, devoted part of its 24 March 1962 issue to a symposium of articles on this subject, with an introduction by Harold Urey.

Urey made the startling suggestion that the life forms observed under the microscope of Nagy and Claus lived on the earth but had reached the New York laboratory via the moon. He cited his proposal that most of the stony meteorites are fragments of the moon kicked free by the impacts of larger bodies and pointed out that the meteorite 'fossils' resembled water algae. Might it not be, he wrote, that the splash produced when a huge meteorite hit the earth carried some of these algae to the moon, where they soaked into the porous rocks and remained until another meteorite kicked them back again? It is thought that the moon was once far closer to the earth than today.

Urey pointed out that several other scientists had proposed that we may find microscopic life forms on the moon. In 1959 Carl Sagan, then at Yerkes Observatory, read before the National Academy of Sciences two papers bearing on the subject. The moon, he said, must once have had an atmosphere similar to the original atmosphere of the earth. If not replenished, it would have diffused off into space in about 1,000 years, but presumably the lunar crust steadily gives off gases, and this may have been sufficient, he said, to stretch the lifetime of the lunar atmosphere to 100 million years. This would have been long enough for the synthesis of a considerable quantity of organic compounds in the manner postulated for the primitive earth. If life never emerged from this process and the organic compounds were protected, by meteoritic dust, from violent temperature changes and solar radiation, then these early compounds, extinct on earth, await the first scientists to reach the moon.

However, Sagan also raised the possibility that there might be a habitable region of the moon, far enough below the surface to be uniform in temperature, warmed by underground radioactivity and moistened by water released from the rocks. If life evolved rapidly when the moon still had an atmosphere, it might still lie hidden beneath the surface, he said. Although this was 'admittedly very speculative', he added, the idea that there may be something akin to life on the moon 'must not be dismissed in as cavalier a manner as it has been in the past.'

The burden of his talk was a warning against contamination of the moon by non-sterile space craft. Bacteria from the earth, he said, might find abundant compounds on which to feed, with a consequent 'biological explosion' that would quickly destroy the precious, primitive substances. Or the microscopic invaders from earth might find living natives and devour them before they could be discovered and studied. Such possibilities, he said, are remote, but sufficiently real to make it imperative to sterilize space vehicles destined for the moon. To the dismay of some biologists, it was later decided to relax moon sterilization procedures in the interest of economy (and, presumably, to improve the chances of reaching the moon before the Russians). It was argued that any bacteria surviving the trip would almost certainly be killed by ultra-violet radiation or other stresses on the unprotected lunar surface, whereas it was generally agreed that contamination of Mars by organisms protected within a spaceship would be intolerable.

An even more 'far-out' hypothesis discussed by Urey in his introduction to the *Nature* symposium was that advanced by John J. Gilvarry, a British-born scientist who had taught at Princeton University and who, at the time he sent his proposal to *Nature* in 1960, was with the Research Laboratories of the Allis Chalmers Manufacturing Company in Milwaukee, Wisconsin.

Gilvarry assumed that the moon, to begin with, had as much water as the earth or any other body of the solar system. If the moon had the highlands that we see there today, they would have been surrounded by seas two kilometres deep. The moon should have kept its oceans at least a billion years, he said. For various reasons, including its thin atmosphere, there would have been little rain, accounting for the fact that we see no river valleys, but he proposed that close examination of the lunar surface should show the branched patterns of moderate erosion.

He argued that a number of lunar craters, such as the huge bowl of Mare Imbrium, must have been formed by impact explosions that occurred under water. This, in fact, was his chief argument and was based on the study of craters on earth, including those produced by atomic-bomb tests in Nevada. The smooth floors of the 'seas' visible on the moon are, in fact, oceanic sediments, he said, and are dark with the vestiges of 'a primitive form of life' that existed in the vanished oceans.

Urey's reply was based on the generally accepted view that, in its early infancy, the earth had no atmosphere or oceans, its envelope of gases having been swept away. The present air and surface water thus must have come slowly from within the earth through volcanic action and the breakdown of surface materials. There is no reason to believe, Urey said, that the moon exuded its water in a great rush at the start. More probably it, too, has done so at a fairly uniform rate throughout its history, but the moon, with its weak gravity, was unable to retain the water and other gases; they diffused away as fast as they were liberated.

Another proposed source of biological material on the moon was 'astroplankton'. This was an elaboration, by J. B. S. Haldane, of the panspermia hypothesis. Ordinary plankton is the drifting life of the sea, whereas astroplankton is dormant material flying every which way through space.

In 1954, three years before the first sputnik, Haldane had written: 'One of the earliest parties to land on the moon should be able to look for astroplankton, that is to say, spores and the like, in the dust from an area of the moon which is never exposed to sunlight.' If the material were for ever shaded, he felt, it would not be altered by prolonged exposure to solar radiation.

Haldane discussed the astroplankton idea as a second-choice explanation for the origin of life. In a reference to the steady-state theory of the universe, promulgated two years earlier by Hermann Bondi, Fred Hoyle and Thomas Gold, he said life itself may have had no origin:

> The Universe may have had no beginning [he wrote]. I do not think it had. Further, if it had no beginning, some parts of it may at all times have been in the condition of the parts which we know, and have included some niches where life was possible. . . . On such a view life is co-eternal with matter.

Not only could life spores be carried from one part of the universe to another by light pressure, he said. It was even possible that they were 'launched into space by intelligent beings'. This may seem far-fetched, he admitted, but if chemical evolution finally proves unlikely, he thought such ideas would have to be taken seriously.

Anders, while leading the attack on those who thought they had found fossils in meteorites, sought also to explore whether Haldane's proposal could explain the presence of spores in meteorites knocked off the moon. He examined, in particular, how many spores would be likely to reach the moon from the earth and other planets. His calculations indicated that, for every spore from Mars, there would be 3,300,000 spores arriving from the earth. Impacts of meteorites and comets on the earth would, as proposed by Urey, throw additional material up to the moon, he said. However, spores from beyond the solar system would be so rare as to be undetectable. Not that this rules out the panspermia hypothesis, he added, particularly if living earth spores are found on the moon, for it would take only one viable spore to account for life on our planet. 'We may all be descendants of this spore,' he suggested provocatively.

Among contributions to the 1962 symposium in *Nature* was an analysis of the reports of Nagy and his New York associates, written by Anders with two of his colleagues at the University of Chicago: Frank Fitch of the Department of Pathology and Henry P. Schwarcz of the Enrico Fermi Institute for Nuclear Studies. They said that in specimens of the Ivuna and Orgueil meteorites they, too, had found the spherical and oval forms described by the New York group. 'Meteorites have long been notorious for containing structures resembling fossils,' they said. They cited the discredited report published by Otto Hahn at Tübingen in 1880 describing a great variety of 'organisms' that he had found in chondritic meteorites. While the work of Claus and Nagy had been done 'with much greater care and competence', the Chicago group said, 'the decision whether a certain form is of biological or inorganic origin is . . . quite subjective.' In fact, the Chicagoans asserted that the rounded particles were striking similar to supercooled droplets of sulphur and hydrocarbons produced by experiments in their laboratory.

Nagy, Claus and Hennessy, far from capitulating to the objections of the Chicago group, said the latter's droplets of sulphur and hydrocarbon bore no relationship to the objects of their own study. They produced additional pictures of lifelike objects extracted from Orgueil and Ivuna and reported finding further samples in two other carbonaceous chondrites: Alais, the first of this kind to be discovered, and Tonk, which had fallen in India in 1911. A specimen of the most lifelike form of all – the hexagonal Type Five – was obtained from Tonk.

In a commentary forming part of the symposium, Bernal admitted to being impressed by the slides shown him by Claus. He felt the claims of the New York group merited serious consideration, despite the questions that they raised. What troubled him in particular was the suggestion that life arising elsewhere would display the same chemistry as life on earth. He termed the alternatives equally unlikely: Either life, disregarding all other possible paths of development, always evolves the same chemical structure no matter where it arises, which he considered 'inherently improbable', or life on earth and all life elsewhere has a common ancestry, which, he said, 'strains the imagination'. Perhaps, he added, Haldane was right and our spark of life came from elsewhere in the galaxy, 'or indeed from other galaxies'.

Bernal pointed out that the impending examination of the moon's surface may help resolve these basic questions. However, to settle the dispute between the New York and Chicago groups, he made a proposal reminiscent of the efforts in Paris to resolve the controversy between Pasteur and Pouchet regarding spontaneous generation. The argument, he said, 'clearly requires for its resolution careful comparisons by a panel of impartial experts'.

By now the debate had developed to the point where the New York Academy of Sciences, with Nagy as a prime mover, had decided to call another meeting on the subject to be held on 1 May 1962. Most of the leading protagonists came and Harold Urey served as chairman. New supporters of the New York group came forward, including A. Papp of the Department of Paleontology at the University of Vienna and Pierre Bourrelly of the Department of Cryptogamic Botany at the Museum of Natural History in Paris. The latter said the objects in the New York slides were definitely organisms and were not contaminants, although he found them more like earth forms than one would expect in life that had evolved elsewhere. Papp felt their resemblance to aquatic life on earth did not mean they had been aquatic in their distant home.

Sidney Fox, who had cooked a mixture of amino acids into a protein-like material that formed into tiny, bacteria-like spheres, suggested the 'organized elements' might, in fact, be such spheres. Paul Tasch of the Department of Geology at the University of Wichita, who, experimentally, had explored the possibility of contamination, said everyone might be right: some of the forms could be terrestrial contaminants; some could be fossil remains of extraterrestrial organisms; some could be Fox's spheres.

One of the critical aspects of the debate concerned a staining procedure used to detect the presence of DNA (deoxyribonucleic acid). The New York group had, in this way, found what seemed DNA-like material in some of the suspected fossils – strong evidence for their biological origin.

In this procedure, the so-called Feulgen test, the sample is treated with hydrochloric acid, which 'loosens' the DNA molecules, exposing their deoxyribose phosphate groups. If a substance known as Schiff's reagent is then added, a characteristic red colour is produced.

The New Yorkers had found that many of their Type One and Type Two forms stained pink in a manner typical of certain bacteria and blue algae whose nuclear material (including DNA) is spread throughout the cell.

Anders and Fitch of the University of Chicago argued that this did not necessarily show the presence of DNA or a related substance. The Feulgen test is effective as an indicator of DNA, they said, only after the elimination of other substances that might also stain red when exposed to Schiff's reagent. Anders and Fitch applied the test to samples of rat spleen, the Orgueil meteorite and kimberlite, diamond-bearing rock from South Africa that was formed more than 100 miles below the earth's surface and has a mineral composition somewhat like that of the Orgueil meteorite.

The results of all three tests were positive, they said. Since the material that stained red in the kimberlite was clearly not of biologic origin, there was no certainty that the meteorite sample had fossil material either. They did similar tests on starch grains and obtained a pink colour very much like that obtained by the New Yorkers in their Types One and Two. They concluded that some of the meteorite objects may have been grains of starch and that those with a more lifelike appearance were probably pollen grains that had infiltrated the specimen after its arrival on earth.

By now the debate was not entirely academic, Claus, Nagy and Dominic L. Europa of the Department of Pathology at Bellevue Hospital in New York were clearly piqued at the suggestion that they did not know how properly to interpret Feulgen staining. Of the Chicago group they said: 'Unfortunately, these authors apparently never carried out real Feulgen staining on their starches, or, if so, they did not trouble to mention their undoubtedly negative results.'

They were also stung by the implication that, during preparation, their specimens had become needlessly contaminated. In examining slides of meteorite material prepared in the Chicago laboratories, they also found fragments of earth life, showing, they reported acidly, 'that experience is essential in preparing uncontaminated samples . . .'. Later, in the published version of these papers, the sharpest of the comments were deleted.

In summary they claimed they had, by now, studied 400 microscopic preparations of meteorite and related materials, disclosing 30 distinct types of organized elements. More, they said, had been found by other researchers. None were identical to earth species, though some were similar. 'Full proof' of their extraterrestrial origin is lacking, they stated, 'but the indications seem to be strong.'

One participant in the academy discussions, Philip Morrison of Cornell University, was a physicist best known for his work on cosmic rays and other subjects remote from the question of life in meteorites. However, his intense interest in the possibility of communicating with intelligent beings elsewhere had led him to ponder the evidence presented by Claus and Nagy. He came forth, in a contribution to *Science*, with the startling suggestion that the 'life' forms in meteorites might actually be 'snowflakes', formed in space as aggregates of carbonaceous materials. He cited the striking manner in which true snowflakes are formed, each in a well-ordered design, yet in a great variety of patterns. Since all the flakes form in similar environments from a very simple substance – water vapour – it is puzzling that all six arms of a flake grow in the same design. One possibility, Morrison pointed out, is that instructions are somehow transferred acoustically from one arm to another. In a somewhat similar manner, he said, long-chain organic molecules could have built into the variegated forms observed by Claus and Nagy. Such 'snowflakes', he said, may constitute an important stage in the evolution of living cells.

A year after the New York Academy of Sciences meeting, a severe blow was dealt to Claus, Nagy and their co-workers. As noted earlier, of all the 'organized elements' that they had found, by far the most convincing were those of the six-sided Type Five variety. They were comparatively rare. Two, plus a few fragments, had been found in Orgueil and four in Tonk. When stained by the Gridley method (used particularly in microscopic studies of fungi) their features came into clear relief: tufted protuberances from alternate faces of their hexagonal structure as well as other symmetrical characteristics.

On 7 June 1963 Fitch and Anders of the Chicago group fired their salvo in a report published in *Science*. They had tested Gridley staining on ragweed pollen and found that the grains thus treated were indistinguishable from the Type Five elements. The reason, as some had suspected, was that the staining distorted the grains.

This convinced most neutral observers that the Type Five elements were, in fact, pollen grains that had entered the samples while on earth. Such grains are a notorious contaminant of New York City air during the hay fever season. Furthermore, word got around that Anders had found something even more remarkable in a purported Orgueil specimen newly furnished by the museum at Mountauban, near the site of the original fall. Embedded within it were several suspicious-looking pods, one of them attached to a stem that ran directly through the specimen. The plant proved to be a native of southern France, and those sceptical of the Orgueil 'fossils' rejoiced in the fact that the plant was identified as *Juncus conglomeratus*. In pronouncing the name they laid special emphasis on its first syllable.

In truth, however, it had no bearing on the dispute. The specimen had apparently been dissolved, the stem and pods inserted, and the material then allowed to cement itself back together. The museum officials expressed dismay. The specimen, they said, had lain undisturbed under glass since shortly after its fall in 1864. Was this the work of some nineteenth-century prankster who hoped the specimen would fall into the hands of Louis Pasteur or other participants in the debate on spontaneous generation? If so, he little anticipated that, instead, it would figure in quite a different controversy a century later.

Despite their Type Five setback, the New York group now had a powerful ally in Harold Urey. Not that Urey was convinced that the meteorite objects were fossils, but he felt the proposals of Claus, Nagy and the others were not being given a fair trial. 'It should be realized,' he said after the New York Academy meeting, 'that enthusiastic people may misclassify artifacts of one kind or another, or may mistake a contaminant for an indigenous form; in fact, it would be surprising if this did not occur in such a study. On the other hand, enthusiastic critics also make mistakes.'

Urey helped Nagy gain access to some of the newest and most sophisticated analytic devices of the day. One of these was the 'electron probe microanalyser' on Urey's own campus at the University of California in La Jolla. Within limits such a device can determine the composition of objects even smaller than the suspected meteorite fossils. In fact, it can scan across them, revealing compositional variations in their structure.

The microanalyser fires an extremely narrow beam of electrons at the target, causing the latter to emit X-rays. The resulting X-ray spectrum discloses the target composition in terms of the heavier elements. The electron bombardment of objects extracted from Orgueil showed a number of them to contain iron with smaller amounts of chlorine and nickel. The metal, in some cases, was concentrated in the wall enveloping the object. The report on these findings, published in *Nature* with Urey as a co-author, said the content of these objects was compatible with fossilization in which cells are invaded by durable material. They were clearly not earthly pollen grains. However, Anders argued that they could be mineral grains rather than fossils.

Nagy's next step was to visit the Karolinska Institute in Stockholm, part of the Medical Nobel Institute, where there was a 'Universal ultramicrospectrograph'. This instrument, like the one at La Jolla, could study objects as small as one micron (a millionth of a metre) in diameter. It obtained ultra-violet absorption spectra of the Orgueil specimens after the latter had been purged of acid-soluble metallic constituents. The results were described in *Nature* as being at least consistent with the presence of nucleic acids and proteins.

On 14 and 15 May 1964, when a symposium was held at La Jolla to mark the centennial of the Orgueil fall, two new discoveries were reported that Urey felt should persuade the scientific world to take seriously the arguments of Nagy and his colleagues. One was the presence in the meteorite of porphyrins; the other was the detection of 'optical activity'.

Such activity – the ability to rotate light waves – derives from the peculiar asymmetry of certain substances synthesized by all living organisms. Its significance as a phenomenon peculiar to life was recognized by Pasteur in his early work on tartaric acid, and, as noted in Chapter 8, led him to believe that the chemistry of life could not have evolved from inorganic substances.

The materials that display optical activity, such as sugars and amino acids, can form molecules in two alternate configurations that are mirror images of one another. The two forms are identical chemically, and, when formed in equal proportions by non-biological processes (such as the origin-of-life laboratory experiments) have no effect on light waves. However, the amino acids from which all the proteins of life are formed represent only one of the two mirror images; consequently, when light waves pass through such material, they are rotated to the left.

To observe such rotation the light beam must be plane-polarized. That is, the light waves must be vibrating in one direction, like the waves in a horizontal rope that is being shaken up and down. It is then possible to see if the wave motion has been rotated in one direction or the other. Whereas the amino acids in protein always twist light to the left, the sugars in the long molecules that carry heredity (ribose and deoxyribose) rotate it to the right. There seems to be no natural process, apart from life, that produces such asymmetrical substances; hence optical activity, as it is called, has long been considered an indicator of life-produced materials, even in the most ancient sedimentary rocks.

Several investigators had tried to detect optical activity in the carbonaceous chondrites, including Mueller in his early work on

Cold Bokkeveld, and Briggs and his colleagues in New Zealand, in their study of Mokoia. However, in the search instituted by Nagy's group, as described in its report to *Nature*, three identically prepared extracts from three different Orgueil specimens were measured by three independent investigators at three laboratories. All showed rotation of light waves to the left. Parallel tests on dust from the Montauban museum, on museum wax and on soil and pollen, as examples of possible contaminants, all rotated light to the right. Material from a different type of meteorite (the Bruderheim) showed no activity at all.

An interesting proposal by Carl Sagan is that bacteria entered the meteorite, after its fall, and ate organic compounds native to the specimen – compounds that had been synthesized inorganically in space. The bacteria ate only one of the two mirror images of such molecules leaving a residue that was one-sided – that is, optically active.

The other piece of evidence, regarding porphyrins, came from G. W. Hodgson and B. L. Baker of the Research Council of Alberta, in Edmonton. As noted in the discussion of chemical evolution (Chapter 9) the porphyrins and their cousins, the chlorins, are among the most universal, vital and characteristic chemicals of life. Because of their special light-absorbing qualities they constitute the pigments in many living substances. Hemin, a porphyrin, makes the blood red; chlorophyll, a chlorin, gives leaves their green colour. The chlorins differ from their porphyrin counterparts in having two extra hydrogen atoms. When left for long periods in sediments, the chlorins apparently break down into porphyrins, so that old sediments can be identified by their lack of chlorins. In the debris of recent life the chlorins predominate.

Hodgson and Baker reasoned that, if both chlorins and porphyrins were found in Orgueil samples, they might represent contaminants. But if only porphyrins were found, as in ancient sediments, the material was more likely of fossil origin. They studied a variety of samples, including six from Orgueil, material from non-carbonaceous meteorites, various dusts and known porphyrin and chlorin samples as a basis for comparison. The dust included vacuum-cleaner sweepings, long-accumulated dust from the top of a home furnace and undisturbed dust from under the organ of an old church.

The non-carbonaceous meteorites showed no reliable evidence of either substance. The dusts were more rich in chlorins than in porphyrins, whereas the Orgueil samples contained virtually no chlorins. Their porphyrin content (vanadyl porphyrin) seemed typical of that in ancient sediments on earth and in petroleum.

Both this report and that of Nagy's group, while arguing for extraterrestrial origin of these substances, conceded that the possibility of contamination had not been totally eliminated. The opposition argued that, since porphyrins are thought to have evolved on earth as a prelude to the origin of life, they could also have evolved elsewhere.

One of the few areas in which the rival groups are in partial agreement concerns the environment of the body from which the Orgueil meteorite came. A peculiarity of the carbonaceous chondrites, recognized early in their study, is the enormous amount of water chemically wrapped up in their material. Orgueil and Ivuna are about one-fifth water. Anders and Eugene R. DuFresne estimated that the hydrated silicates in Orgueil must have been exposed to liquid water for at least 1,000 years. Yet liquid water is a rare commodity in the solar system. Anders proposed that there may have been an ice layer beneath the surface of an asteroid that trapped liquid water beneath it. The trapped water would have been kept warm by radioactivity in the surrounding rocks and he conceded that, conceivably, life might evolve in such a subterranean environment. However, he rejected the arguments of the New York group as evidence for this.

Nagy and his colleagues also discussed the possibility of buried life, citing Sagan's arguments for organic matter beneath the surface of the moon. They said the possibility remained, as well, 'that the parent body was of sufficient size to hold an atmosphere and thus bodies of water.' However, as Urey has pointed out, the carbonaceous meteorites bear no resemblance to the sorted, layered material in sedimentary rocks laid down beneath bodies of water on earth.

The Orgueil debate is a classic example of a scientific discussion become personal, emotional and enmeshed with professional pride. The talents and ingenuity of participants have been directed toward proving their case, rather than seeking out the truth. They have thus demonstrated that they are human, but the wonderful self-discipline and objectivity that we call pure science has suffered.

The inconclusiveness of the discussion also reflects the inadequacy of our analytic methods. The Orgueil meteorite has probably been more elaborately studied than any other chunk of material on earth. Yet to those who marvel at the sophistication of present-day techniques, it is sobering to observe that the complex and varied components of the specimen defy precise definition. A pity, for, as Harold Urey has said, 'If it can be shown that these hydrocarbons and the "organized elements" are the residue of living organisms indigenous to the carbonaceous chondrites, this would be the most interesting and indeed astounding fact of all scientific study in recent years.'

12

Is There Life on Mars?

No object in the heavens has been the subject of such bitter controversy in recent years as the planet Mars. The claims of those who thought they saw signs of life there have produced counter-statements by distinguished scientists who said it was unquestionably a dead planet – a viewpoint reinforced by the Mariner photographs of 1965. Yet, on the eve of space flights that will ultimately tell us the truth about Mars, a remarkable diversity of ideas has flourished. Thus in 1962 an American scientist suggested that flashes observed on Mars, followed by the appearance of small clouds, might be atomic bomb explosions. This admittedly far-fetched idea was set forth in so respectable a journal as *Science.* A Soviet astrophysicist said the moons of Mars may be artificial, launched by an extinct civilization. Another Russian has proposed that Mars may furnish us with 'cattle' of extraordinary usefulness.

While the possibility of life in other worlds has been discussed since ancient times, attention until recently was focused on a variety of bodies. In fact only in 1963, when evidence that Venus is oven-hot was confirmed to the satisfaction of almost everyone, did the search for life in the solar system narrow to Mars.

Discovery of the essential features of that planet began in the seventeenth century, when Huygens made drawings that showed a white cap on one of its poles. Others observed that there are periodic changes in its dark markings, and in the eighteenth century Sir William Herschel noted that these included alterations of colour. The white polar caps were of snow and ice, like those on earth, he said, and he argued that the 'inhabitants' of Mars enjoyed an environment much like our own. Herschel observed the planet on its close approaches, or 'oppositions', of 1777, 1779, 1781 and 1783. Of all the planets Mars is the most easily observed. Its orbit around the sun is the next outward from the earth, but because both orbits are elliptical, the distance between the earth and Mars, at opposition, is highly variable. The 'close oppositions', when Mars comes within 35,000,000 miles, occur every fifteen to seventeen years, the next being in 1971. Clouds completely hide the surface of Venus and, because its orbit lies inside that of the earth, its near side is dark at its closest approach.

It was two observations in the year 1877 that drew the attention of the world to Mars. One was by Giovanni Virginio Schiaparelli of Milan Observatory in Italy, who saw that he called 'canali'. The other was the discovery by the United States Naval Observatory of two Martian moons, one of them with a rather peculiar orbit.

Schiaparelli's 'canali' would probably have best been translated as 'channels', but instead they became known as canals, implying construction by Martian engineers organized within some superior global society. Schiaparelli himself was non-committal: 'Their singular aspect, and their being drawn with absolute geometrical precision, as if they were the work of rule or compass,' he wrote, 'has led some to see in them the work of intelligent beings, inhabitants of the planet. I am very careful not to combat this supposition, which includes nothing impossible.' Nevertheless, he said, nature often manifests symmetry without the intervention of intelligence, as in crystals, spiral sea-shells, and so forth. 'The perfect spheroids of the heavenly bodies and the ring of Saturn were not constructed in a turning lathe, and not with compasses has Iris described within the clouds her beautiful and regular arch.'

Soon thereafter word reached a Bostonian named Percival Lowell that Schiaparelli's eyes were failing, and Lowell decided to pick up the Italian's torch. He was the brother of Amy Lowell, the poetess, and of Abbott Lawrence Lowell, for twenty-four years president of Harvard. Percival was primarily an orientalist, having lived in Japan and Korea, written extensively on the Far East, and served as foreign secretary to the Korean Special Mission to the United States. Now, in his thirties, he became caught in the excitement of Schiaparelli's observations and decided to become an astronomer.

In 1894 Mars was to pass in close opposition, and Lowell went to Arizona to observe it through the clear, dry air of that region. 'A steady atmosphere is essential to the study of planetary detail,' he wrote, 'size of instrument being a very secondary matter. A large instrument in poor air will not begin to show what a smaller one in good air will. When this is recognized, as it eventually will be, it will become the fashion to put up observatories where they may see rather than be seen.'

His prediction has been dramatically fulfilled in Arizona with the establishment of what may soon be the most extensive complex of observatories in the world. The fruit of his own efforts, the Lowell Observatory at Flagstaff, was the first well-equipped station devoted primarily to study of the planets.

Lowell's observation of the oppositions in 1894 and 1905 added enormously to the knowledge of Mars. He calculated that its atmosphere was only half as dense as that on the summit of Mount Everest, and that Mars 'is very badly off for water', depending for its meagre supply on the melting of its polar ice in spring. He and William H. Pickering, who had studied Mars from Harvard's field station at Arequipa, Peru, agreed that the dark areas of Mars may once have been seas, but are now regions of vegetation. Both men described in detail the canals reported by Schiaparelli. They seemed to extend from the dark areas into the reddish desertlike regions and also radiated from dark spots, or 'oases', like the spokes of a wheel. To be visible from the earth they would have to be many miles wide, and hence it was proposed that what was seen were bands of irrigated land, rather than the canals themselves.

All this, of course [Lowell wrote], may be a set of coincidences, signifying nothing; but the probability points the other way. ... Irrigation, unscientifically conducted, would not give us such truly wonderful mathematical fitness in the several parts of the whole as we there behold. A mind of no mean order would seem to have presided over the system we see – a mind certainly of considerably more comprehensiveness than that which presides over the various departments of our own public works. Party politics, at all events, have had no part in them; for the system is planet-wide. ... Certainly what we see hints at the existence of beings who are in advance of, not behind us, in the journey of life.

The argument of Lowell was that the peculiar features of Mars – its seasonal changes and its markings – were most easily explained by the presence of life, even though other explanations were possible. In essence this is still the reasoning of those who believe there is some form of life there. Yet in Lowell's day, as now, there were vehement doubters. Astronomers of repute looked at Mars and said they could see no canals at all. Alfred Russel Wallace, who independently devised a Darwinian theory of evolution, said flatly that Mars is lifeless and that 'water-vapour cannot exist' there, a statement shown to be incorrect in 1963 when water was finally detected in the Martian atmosphere.

Another sceptic was Svante Arrhenius, author of the concept of panspermia, who explained the changing colours of Mars in terms of hygroscopic, or water-hungry, salts deposited on the bottoms of shallow lakes. These lakes, he said, dry out in winter and the moisture is deposited on the winter pole as ice crystals. When these melt, in the spring, the air-borne moisture is absorbed by the salts which thus are darkened. This would account for a peculiar feature of the Martian spring: namely, that the darkening moves southward from the pole at about twenty-eight miles per day instead of in the other direction. It is as though spring first came to Labrador and migrated southward to Florida. Others have proposed that the white material on the poles might be dry ice (frozen carbon dioxide) or nitrogen tetroxide. Advocates of the latter view proposed that the Mars air is poisoned with nitrogen oxides, ruling out any form of life at all. However, spectra obtained by three observatories seem to have eliminated this hypothesis and it is generally agreed, from studies of light reflected from the polar caps, that they are coated with ordinary frost.

The reason for the violent differences of opinion about the canals rested with the difficulty of photographing detailed features on a planet, or even on the moon. When one looks at such a body through a powerful telescope it is like gazing up through running water. For fleeting instants fragments of the picture seem to come through clearly, but before one is sure of them they become distorted or dissolve because of movements within the earth's atmosphere. Hence, in photographs, Mars appears frustratingly fuzzy. However, the human eye is an observing instrument without match, for not only is it extraordinarily sensitive, but it can store the fleeting bits of the picture that it sees and fit them to other fragments to produce an integrated picture. The strength of this system – the human brain – is also its weakness, for it is subjective. Unwitting eagerness to 'see' certain features predisposes the viewer to string glimpses together into patterns that may not actually be there.

The photographs of Mars transmitted to earth by Mariner IV after it flew past that planet on 14 July 1965 show no striking straight-line features, although the pictures spanned several charted 'canals'. Rather, they reveal a landscape so pitted with craters, old and young, that it could not have been subject to water erosion, at least not for a very long time. As noted earlier, the fact that the earth from time to time has been hit by large objects is evident from a few impact craters, but mountain-building processes have ploughed under the older scars, water erosion has worn them down and sedimentation has buried them. On the moon, where such processes are absent, even extremely ancient craters still show up dimly. The Mariner photographs implied that on Mars, as well, water action and mountain-building have long been absent. However only a few of the twenty-one full images received on earth show the surface with sufficient clarity to distinguish features two or three miles wide. Also the entire area scanned was less than one per cent of the planet. Hence, the pictures do not throw much light on the large-scale surface features. In particular they do not explain what distinguishes the light from the dark areas.

It was once thought the dark areas might be like forests on earth, but temperature observations made this seem doubtful. During the Martian day, which is almost exactly the same length as that on earth, dark regions near the equator become 15 degrees hotter than nearby reddish areas. If they were covered with dense vegetation, one would expect them to be cooler. Furthermore, they warm in the morning and cool in the evening just as fast as the seeming deserts, reaching their maximum equatorial temperature of 90 or 100 degrees a half hour after midday. Our own vegetated regions are much slower to change temperature.

Because our orbit lies between that of Mars and the sun, we can never see the midnight portion of that planet's surface, but the rate at which its temperature drops in the evening indicates that even in the tropics it falls as low as 95 degrees below zero. Thus only for a brief period each day is the temperature above freezing and, according to a recent study by NASA scientists, elsewhere on the planet it rarely, if ever, thaws except in nuclei of the dark areas during the annual darkening. In fact, the seasonal disappearance of the polar white cap does not apparently represent melting of the frost, but rather sublimation – direct transition from a solid to a gaseous state. This is said to occur there when the temperature rises to about 95 degrees below zero. Mars is so very cold in part because, on the average, it is 50 million miles farther from the sun than is the earth. The radical day-night variations in temperature are a consequence of its thin atmosphere and its lack of oceans, both of which serve as tempering influences on earth. Mars, with a diameter half that of the earth, has a surface gravity only 38 per cent as strong, and its ability to retain gases is correspondingly less. For a large part of its history, Mars may have had considerable oxygen and water vapour in its air, but these gases slowly diffused into space and probably exist today only because there is a small, but persistent, release of gas from Martian rocks, and perhaps from volcanoes.

The detection of water in the atmosphere of Mars was a triumph of observation technology. It was achieved by the 100-inch telescope on Mount Wilson in California, using the infra-red spectrum of sunlight reflected from the planet as a clue to the amount of water vapour and carbon dioxide in the Martian air. Both these gases absorb certain wavelengths of infra-red light. The difficulty that chiefly baulked earlier attempts to make this observation was the presence of these gases in the earth's own air. It had been impossible to separate the slight effect of the Martian air from the gross effect produced by our own atmosphere, although it had been recognized that this might be overcome by taking advantage of relative motion between the earth and Mars. Such motion, it was thought, might shift the wavelengths of light from Mars sufficiently to separate the spectrum of Martian water from that of earthly water. Not until 1963 did the techniques become sufficiently refined for such an observation.

The distance between the earth and Mars at the time was widening at several miles per second, and this was enough to bring the spectrum of Mars water out into the open. It was calculated that water in the Martian air is from 1,000 to 2,000 times less dense than in that of the earth. On the other hand the carbon dioxide content on Mars was found to be much higher than on our planet.

Similar carbon dioxide values were obtained by Gerard P. Kuiper, head of the Lunar and Planetary Laboratory of the University of Arizona and a veteran observer of Mars. At a 1964 meeting of the American Geophysical Union he put composition of the Martian atmosphere at two parts of nitrogen to one of carbon dioxide, with a little argon and even smaller quantities of water vapour and oxygen. By contrast the air of the earth is 76 per cent nitrogen and 23 per cent oxygen. On the basis of observations at the McDonald Observatory in Fort Davis, Texas, he concluded that air pressure on the surface of Mars is only about 1 per cent that on earth. The Mount Wilson observations led to a pressure estimate almost as low as Kuiper's.

These estimates were discouraging to those designing the vehicles that are to land instruments and ultimately, perhaps, men on Mars. In so thin an atmosphere the usefulness of wings and parachutes would be limited. However, comfort was taken in the knowledge that Mars, with its lesser gravity, holds its atmosphere less snugly than the earth. This, it was assumed, meant that the air of Mars, at very high altitudes, is actually denser than at comparable heights on the earth – a factor which could be exploited in slowing an incoming vehicle. However the Mariner fly-by produced a real shock. As the vehicle passed behind Mars, its radio signals to earth traversed successive layers of the Martian atmosphere before being cut off, for 52 minutes, by the solid mass of the planet. When the signals reappeared on the other side, they again passed through the atmosphere. Analysis of the manner in which they were altered by the Martian air indicated that even the extremely low pressure suggested by the telescopic observations was twice too high. Furthermore, the atmosphere seemed very thin at all heights. The logical explanation of both the ground-based and Mariner observations was an atmosphere composed predominantly of carbon dioxide with only small amounts of nitrogen and argon, plus traces of water vapour and, perhaps, oxygen.

The meagreness of oxygen in the Martian air has been dramatized by Hubertus Strughold, a space-flight enthusiast since his early days in Germany and now Chief Scientist of the Aerospace Medical Division at Brooks Air Force Base in Texas. He points out that visitors to that planet will be unable to build camp-fires, for there is not enough oxygen to support an open flame. More serious for Martian life is the likelihood that there is little or no ozone, the three-atom form of oxygen (as opposed to the oxygen we breathe, whose molecules consist of two atoms). The ozone region of the earth's atmosphere, from 12 to 38 miles aloft, absorbs a large portion of the ultra-violet components of sunlight. Otherwise the world would be uninhabitable for most of its present plants and animals.

However, as noted in Chapter 9, our most remote one-celled ancestors probably were not so sensitive to ultra-violet, a good deal of which may have penetrated the early atmosphere, aiding in synthesis of the pre-life molecules. The early life forms seem to have thrived without oxygen, as do some bacteria today, and, had such an environment persisted, life would have evolved in ways capable of coping with it.

It has also been proposed that the blue haze of Mars absorbs some of the ultra-violet. This haze becomes evident when one photographs Mars in blue light. The surface features come out dim or invisible, whereas in red light they are comparatively clear. Kuiper has reported evidence that the haze is formed of ice crystals like those in cirrus clouds on earth, but even smaller. The haze is so thin, he says, that, if melted, it would hardly be enough to moisten the planet's surface.

A further difficulty confronting life on Mars is the intensity of the cosmic rays reaching its surface. As noted in the discussion of meteorites, these rays consist of extremely high-energy particles coming both from the sun and the regions beyond. They rain on the earth from all directions, but are diverted toward the poles by the earth's magnetism. Most of them strike particles in the atmosphere, losing their fearful speed and energy before they reach the ground. The Mariner fly-by led to two discoveries that bear on the intensity of cosmic rays to which organisms on Mars would be exposed. One was the thinness of the atmosphere already noted. The air of Mars apparently offers almost no protection. The other finding was that Mars has no magnetic shield like that which envelops the earth. This is apparently because Mars lacks a liquid core to generate such a magnetic field. The absence of a Martian field was confirmed by a variety of instruments on board the spacecraft.

At an international space science symposium held in Warsaw in 1963, Herman Yagoda, a specialist with the United States Air Force, estimated, from knowledge of the Martian atmosphere then available, that after major eruptions on the sun someone on Mars would receive 1,000 to 10,000 times more radiation than someone on the earth.

The impediments to life on Mars thus seem formidable: low temperature, little liquid water or gaseous oxygen, severe radiation from both ultra-violet light and cosmic rays. Yet supporters of the idea that life may exist there are not discouraged. They concede that life could probably not have evolved on the planet in its present state, particularly with few, if any, lakes or oceans; but they point out that Mars, to begin with, probably had as much water as the earth. If it remained long enough for life to emerge, then this life could have adapted itself as the environment deteriorated. Some plants on earth produce their own antifreeze and function at sub-freezing temperatures. With most earthly organisms, freezing ruptures cell walls, although some, such as mosses and lichens, have been chilled almost to absolute zero and survived.

At the Warsaw symposium, Richard S. Young of NASA told of experiments in which micro-organisms, including one form collected on the rim of the Italian volcano Stromboli, thrived despite freeze-thaw temperature fluctuations designed to simulate those on Mars. The organisms were of the type that can live without oxygen. The results, he said, support the possibility – but not the likelihood – of life on Mars and show how real is the danger that earth bacteria, transported by space vehicle, might spread over that planet, with tragic consequences.

Suppose, as stated by Carl Sagan in a Voice of America lecture, that micro-organisms very like some on earth are found on Mars. 'Do we conclude,' he asked, 'that similar forms have developed independently on the two planets? Or that Mars and Earth had some common biological contact in the distant past? Or that terrestrial spacecraft inadvertently deposited the organisms on a previous mission?' If, in fact, such contamination had taken place, he said, it would be 'a major scientific disaster'.

As noted by J. B. S. Haldane before space travel became a reality, contamination is a two-way threat. It is not inconceivable that some micro-organisms on Mars may be lethal to man or to other earthly life forms. Great care will have to be exercised to make sure that visitors returning from Mars do not bring with them organisms against which we have no defences. An interplanetary quarantine will be necessary.

Since the Warsaw symposium, Young has continued his experiments with bacteria under Martian conditions and has found that those surviving ordinary sterilization procedures cannot endure the freeze-thaw cycle postulated for the climate of Mars. This, he points out, does not eliminate the need for rigorous sterilization of spaceships bound for that planet, since it is imperative that they do not carry tiny stowaways capable, after the landing, of invading the devices designed to detect microscopic Martian life.

The problem of whether or not Mars is vegetated and, if so, by what sort of plants, was for many years the passionate concern of Gavril Adrianovich Tikhov at the observatory in Alma Ata, capital of the Kazakh Soviet Socialist Republic. His work on the planet began during the opposition of 1909, when he took some 1,000 photographs of Mars through different filters. He later studied plants growing at various elevations and latitudes in Central Asia to see if, in environments most like that of Mars, their light-absorbing properties approximated to what he saw on that planet. He concluded that the plants of Mars did, to some degree, resemble the alpine or arctic flora of the earth and pointed out that even in the vicinity of Verkhoyansk and

Oimyakon, the coldest region of the Northern Hemisphere, there are 200 species of plant. If they could grow there, he argued, why should there not be plants on Mars? He made the remarkable proposal that, whereas the dominant colour of earth plants is green, that of the Martian flora is blue and the plants of Venus are orange, the colour being determined in each case by the temperature and radiation factors of the environment. The geocentrism, or earth-centred point of view, that has inhibited the philosophies of men for thousands of years, he said, has also impeded speculation on how plants might evolve and adapt themselves to very different environments from our own.

For a time, so far as Western scientists were concerned, it was not the observations of Tikhov but those of William M. Sinton, an American, that most strongly suggested the presence of life on Mars. His findings implied that, if there were not earth-like forests, at least there were primitive plants. His first observations, made at Harvard during the opposition of 1956, searched the infra-red light of Mars for bands in the spectrum characteristically absorbed by hydrocarbons, such as those in plants. These bands lie near a wavelength of 3·5 microns. (The micron is one millionth of a metre.) With the Harvard telescope he was unable to sight on individual dark areas and so, at the opposition of 1958, he used the giant reflector on Mount Palomar. When it was aimed at Syrtis Major, one of the dark areas, he observed an absorption spectrum which, he reported, 'fits very closely . . . that of organic compounds and particularly that of plants.' One band, at 3·67 microns, was a puzzle at first, but he found that an alga, *Cladophora*, absorbs light at that wavelength. He speculated that the extent of this absorption on Mars might indicate 'a larger storage of food' in Martian plants. These absorptions were not observed when the great telescope was aimed at the reddish, or 'desert' areas of Mars.

Sinton made further observations with this instrument at the next opposition and in 1963 reported his finding as still 'not a conclusive one for vegetation. It only means that there are organic molecules on the surface. If the bands had not been found, however, it would mean that life was very doubtful.'

Melvin Calvin and his colleagues at the University of California in Berkeley expressed scepticism about Sinton's interpretation, and at the start of 1965 others at Berkeley proposed that 'deuterated water' in the Martian air was a more likely explanation of the absorption. Deuterated water molecules are those in which one or both hydrogen atoms have been replaced by deuterium, the heavy isotope of hydrogen that gives 'heavy water' its extra weight.

With commendable scientific objectivity Sinton got together with the Berkeley group and four months before the 1965 Mariner fly-by they jointly published their conclusions: Two of the absorption bands in question were probably caused by traces of heavy water in the earth's own atmosphere. (A third, at 3·45 microns, remained unexplained.) The seeming origin of the bands in the dark regions of Mars was apparently a by-product of observational difficulties, they said, and perhaps Sinton's earlier association of the bands with organic molecules 'was too strongly suggested'.

As co-author of these words Sinton has, in effect, recanted. However, the reality of the seasonal changes remains. The evidence for some sort of alteration on the surface of Mars in spring has been reinforced by the observations of Audouin Dollfus in France. He has found variations in the polarization of sunlight reflected from Mars at various angles, indicating a springtime increase in the size of particles on the surface. This could either be the burgeoning of microscopic life or merely the effect of moisture.

There are long-term, as well as seasonal, changes in the dark areas. Some vanish after being recorded for decades. Others appear in what had seemed 'deserts'. Such was the case with a dark region about the size of Texas that emerged in 1954.

While there is no question about the darkening of such regions in spring, the true nature of their colour is hard to determine. Contrary to early accounts, they are not green. Some speak of them as 'grey'. Beginning in 1924, Vesto M. Slipher, who, with his brother, E. C. Slipher, continued Lowell's work at Flagstaff, tried to find spectroscopic evidence of chlorophyll, the substance that enables plants on earth to perform photosynthesis. No sign of chlorophyll was evident in his observations or those of others, but this did not necessarily mean it was absent. Some earth plants absorb light at the wavelengths at which chlorophyll strongly reflects it, thus cancelling out the tell-tale effect. Also, as Carl Sagan has pointed out, the absorption spectrum of the chlorophyll in our plants is determined, to a considerable extent, by the side chains of atoms in the molecules as it evolved here. On another planet, its structure might be quite different. It is possible, he says, 'that the green coloration of chlorophyll is largely an historical accident. It is unlikely that the same accident would have occurred on Mars'.

One of the arguments in favour of plants or trees was advanced in 1950 by the Estonian astronomer E. J. Öpik. He pointed out that, from time to time, yellowish clouds move across the planet and come to rest on the dark regions, giving them a dusty hue. It has been assumed that these are dust storms. Yet in a matter of one or two weeks the dark shade reappears as though the objects beneath the dust had shaken it off or grown above it. When Sinton cited this argument at a symposium of the National Academy of Sciences, Harold Urey dryly commented that there was no reason why plants on another planet should not have muscles and be able to dust themselves off.

It is particularly difficult to speculate on what sort of life might exist on Mars since, with little oxygen in the air, even the chemistry of its metabolism would differ basically from that of most earth life. More important, the gross lack of water would seem to place a severe limit on the size of life forms. Kuiper has proposed that any Martian plants or animals must be so small that even a flea would not stumble over them. Yet the seasonable colour changes suggest that the cover of changeable material in the dark areas, though very thin, must be dense. Only in the most verdant regions of our continents would seasonal changes (apart from new-fallen snow) be so dramatically evident.

Has the possibility of intelligent life on Mars been completely ruled out? To many scientists it has. The factors that make the planet seem so bitterly inhospitable to life as we know it make it appear even less likely that the evolutionary process could have flourished enough to produce intelligent creatures. However, if there has been evolution on Mars, it must have followed very different paths from those on earth, particularly since its climate began deteriorating. Hence, at least until the Mariner fly-by, a few 'scientific radicals' refused to dismiss the possibility of intelligence. Notable among these has been Frank B. Salisbury, Professor of Plant Physiology at Colorado State University. In a 1962 article in *Science* he argued that, if there are plants, one would expect to find mobile creatures that feed on them. 'And from there it is but one more step (granted, a big one) to intelligent beings.' We should at least try to keep our minds open, he said, 'so that we could survive the initial shock of encountering them'.

Certain features of Mars, he asserted, 'are most easily understood on the assumption that they are the product of intelligent beings.' He referred, in particular, to the peculiarities of its two satellites and to what have seemed to be eruptions or explosions on its surface. 'Was this volcanic activity,' he asked, 'or are the Martians now engaged in debates about long-term effects of nuclear fall-out?'

He even suggested that, when a robot lands and extends a mechanical arm to grasp a sample of the surface, the results radioed to earth may be perplexing: 'At least I can imagine how I might react if such an apparatus landed in my back yard and started grabbing for my apple tree, the cat, and maybe me!'

The eruptions to which he referred were reported in 1955 by Tsuneo Saheki of Osaka Planetarium, who descibed four occasions (in 1937, 1951 and 1954) when he or one of his countrymen saw a pinpoint of light that scintillated like a star. In the first two cases the light was visible five minutes, whereas in the last it endured only five seconds. After the 1951 flash, a peculiar cloud hovered over the spot. He dismissed the idea of artificial explosions as 'unreasonable' and proposed instead that these were volcanic eruptions.

Saheki's contention that he had seen such eruptions was criticized on the ground that even the most fearsome volcanic events on earth would be hard to see from such a distance. In fact, Carl Sagan argues that, if there are any inhabitants on Mars, they can see little on earth to suggest the presence of life, intelligent or otherwise. The lights of the largest cities could hardly be seen with the equivalent of our most powerful telescopes and, while nuclear explosions could be observed, their flash is so brief that most would not be noticed and those that were would be challenged, since corroboration would be difficult.

If there is volcanic activity, it might produce areas where the environment is comparable to that which existed throughout Mars a long time ago. Such regions, it has been argued, could be comparatively moist. This was suggested in 1962 in a paper written jointly by Carl Sagan and Joshua Lederberg. Four years earlier the latter had won the Nobel Prize in Medicine and Physiology. Their reasoning was that the lack of water in the Martian air did not necessarily mean there was little water beneath the surface. Although the air is dry, a considerable portion of the planet's original quota of water may have frozen underground and thus been preserved from diffusing into space. Lederberg and Sagan suggested that in certain favourable locations, near hot springs, fumaroles, or other volcanic activity, the ground ice would be melted and the planet's internal reservoir of moisture would reach the surface. Hence they suggested that the search for life concentrate on such 'hot spots', if any are found. They could be detected, they said, by space vehicles flying over the night side of Mars, which otherwise is extremely cold.

The Mariner results did not discourage Lederberg and Sagan. The latter, for example, argued that the absence of water-carved valleys in the photographs did not mean that Mars had always lacked rain and oceans. Assuming that Mars, like the earth, is about 4·6 billion years old, it could have enjoyed rain for a billion years or more before its climate dried up. This argument received support from three independent studies published in *Science* on 24 September 1965. They all concluded that the overlapping craters in the Mariner photographs might be no more than a few hundred million years in age. The estimates ranged from less than 300 million to 800 million. Edward Anders, leader of the sceptics in the meteorite controversy, and his colleague, James R. Arnold, concluded that the density of craters 'no longer precludes the possibility that liquid water and a denser atmosphere were present on Mars during the first 3·5 billion years of its history'.

Some of those who most passionately argue the possibility that there is – or has been – life on Mars are men who enjoy being provocative, who like to needle scientific pomposity. Only a few of them, however, have scientific reputations sufficiently impregnable to get away with it. Among these is Iosif Samuilovich Shklovsky, head of the Department of Radio Astronomy at the Sternberg Astronomical Institute in Moscow, who has proposed that the moons of Mars are artificial. In a land that reputedly seeks to control the thoughts of its citizens, Shklovsky is a man of uninhibited intellect. When this writer called on him in 1962 the conversation wandered over a variety of subjects relating to extraterrestrial life. For example, he discussed the possibility that highly evolved civilizations might obtain additional raw materials by blowing up nearby stars. This would be done with a laser producing gamma rays. The laser, an American invention (described more fully in Chapter 15), emits a very narrow and intense beam of light. The effect of such a gamma ray beam on the core of a star, said Shklovsky, might be to trigger a supernova, or cataclysmic stellar explosion of the type that is thought to synthesize the heavier elements. The supercivilization would thus set off supernovae at a safe distance, then harvest the newly created raw materials.

After presenting this idea in some detail, Shklovsky paused, puffing intently at a cigarette held squarely in the centre of his mouth. 'However, I don't think it would work,' he said after a few moments. 'The gamma rays would produce other effects, but I doubt that the star would explode.'

Supernovae are one of Shklovsky's special provinces. His proposal that high-energy cosmic rays originate in such explosions is probably the most universally accepted explanation of those rays.

As he himself has noted, a remarkable feature of the Mars satellites is that they were predicted with considerable accuracy by Jonathan Swift, the English satirist, in his classic, *Travels into Several Remote Nations of the World*, published in 1726 and originally attributed to one Lemuel Gulliver. Among the places that Gulliver visited was Laputa, a strange, rounded island 4 miles wide that travelled through the air, its path controlled by a giant lodestone, or lump of magnetic rock, in its core. The islanders who twisted the lodestone so that its magnetism would interact with that of the earth in various ways were far ahead of the rest of the world in astronomy. In particular, Swift wrote, they had discovered two satellites orbiting Mars at periods of 10 hours and 21·5 hours, respectively (the actual figures are 7·6 hours and 30 hours). Their distances from Mars were three times and five times that planet's diameter, which is considerably more than the actual distances of the moons discovered in 1877.

The squares of their orbital periods are proportional to the cubes of their distances from the centre of Mars, Swift said, 'which evidently shows them to be governed by the same law of gravitation that influences the other heavenly bodies'. The book, now better known as *Gulliver's Travels*, was written during the period when Newton headed the Royal Society and the intellectual world was full of his new ideas on gravity.

The spirit of satire that was abroad produced other flights of fancy. Voltaire in 1752 wrote of Micromegas, a man more than 20 miles high, who came from a planet of Sirius and teamed up with a resident of Saturn, a mere 'dwarf' hardly more than a mile in height, to explore the solar system. Their sojourn on Mars was brief because the planet was too small for them to rest there with comfort, but, like Gulliver, they found it to have two moons.

Voltaire said that, by 'analogy', this was not surprising. He apparently reasoned that, if the earth had one moon and planets beyond Mars had many satellites, Mars should have at least two. In view of the particular distance of Mars from the sun, he said, it 'must be in a very uncomfortable situation, without the benefit of a couple of moons.'

The actual discovery was made by Asaph Hall, who was in charge of the 26-inch telescope at the Naval Observatory in Washington. During the opposition of 1862, the closest of the century, a 48-inch mirror had been used to search for moons near Mars, without success. However, during the next close opposition, on 11 August 1877, Hall observed what he suspected was a moon. The following nights were cloudy, but on 16 August he got another good look and found that the new object had, in fact, moved across the stellar background with its parent body, Mars, showing it to be a satellite of that planet. The following night he found a second moon, even closer to Mars. Hall named them

Deimos and Phobos for the sons of Mars, god of war (or in some versions the horses that drew his chariot).

Since then several peculiarities of these moons have been observed. They fly almost circular orbits, inclined only 1 degree and 1·7 degrees to the planet's equator. This is unusual among naturally orbiting bodies, but more strange is another feature of the orbit of Phobos. It circles the planet roughly three times for every spin of its parent body, being the only one of the thirty-one known moons that does so. All the rest orbit more slowly than the spin of their parents. Furthermore, during the 1940s Bevan P. Sharpless at the Naval Observatory analysed the motion of Phobos and found that its orbit was decaying in a manner difficult to explain. Phobos seemed doomed to destruction in a comparatively short time, geologically speaking. It is now some 3,700 miles above the Martian surface and, if it gets more than about 1,000 miles closer, it will pass within the so-called Roche limit and presumably be torn apart by tidal stresses.

The puzzle, as noted by Shklovsky, was to explain the drag that was apparently nibbling at its inertia. No ordinary factor, such as air drag, magnetic drag, light pressure or tidal friction, seemed adequate to account for the observations of Sharpless. However, the orbital decay could be accounted for if Phobos were hollow, Shklovsky said. It would then be light enough in weight to respond to subtle forms of drag, just as the gigantic balloon satellite Echo I was markedly affected by the pressure of sunlight. The estimated sizes of Phobos and Deimos – 10 and 5 miles in diameter, respectively – had been based on their brightness. They are too small to show up as shapes on the photographic plates of earth-based telescopes, Deimos being the smallest moon of the solar system. Hence calculation of their size has been based on the assumption that they have the same albedo, or reflectivity, as Mars itself. However, if they are spheres of polished metal, they could be far smaller – perhaps only a mile or so in diameter – and still shine as brightly.

When Shklovsky's idea was published, as an interview with the Moscow newspaper *Komsomolskaya Pravda*, it created a sensation. He suggested that, several million years ago, there were oceans on Mars and ample oxygen in its air, but as oxygen and water diffused away the inhabitants were forced to launch satellites and take refuge there. Such a launching would be easier than on earth, since the surface gravity is so much less.

The Soviet Academy of Sciences subsequently asked Shklovsky to write a book in its proper science series, and he chose as his subject extraterrestrial life. When it appeared in 1962, there was a full chapter on the moons of Mars. Perhaps his suggestion was a 'fantastic idea', he wrote, but he questioned whether there was any more reasonable explanation. The book is shortly to be published in English.

Meanwhile, George Wilkins of Britain's Royal Greenwich Observatory had come to the Naval Observatory in Washington to work on the Mars satellites and, in the view of Gerald M. Clemence, the observatory's director, had shown that there were flaws in the calculating methods of Sharpless. Nevertheless, on the eve of his book's publication, Shklovsky insisted that the orbit of Phobos is imperfectly understood and that the question is therefore 'still up in the air'.

While Shklovsky thought in terms of an extinct civilization, others have proposed that the moons may be of recent origin. Frank Salisbury, in his *Science* article, asked why the two satellites were not seen in 1862, when closer and sought by a larger instrument than that with which Asaph Hall discovered them in 1877: 'Should we attribute the failure of 1862 to imperfections in the existing telescopes, or may we imagine that the satellites were launched into orbit between 1862 and 1877?'

While attention – and imagination – have been focused on Mars, the possibility of life elsewhere in the solar system has not been ignored. Venus, our nearest planetary neighbour and earth's twin in size, was long a favourite candidate. Its cloud cover perpetually hides its surface, but it was proposed that the planet might be entirely covered with one great ocean. Any life there, it was said, would have to be marine. However, in 1956 the Naval Research Laboratory in Washington detected radio emissions from Venus suggesting a surface temperature of several hundreds of degrees. A few optimists argued that the emissions might be coming from an upper atmosphere so highly ionized that it is thousands of times richer in electrons than the earth's ionosphere.

However, on 14 December 1962, Mariner II, the first vehicle to send back data from near another planet, scanned Venus on a radio wavelength of 19 millimetres and found the emissions coming from the centre of the planet to be strong, whereas those from its rim were weak. If they were being generated in the upper air, the edges of the planet would have been the strongest source. It was concluded by the experimenters that the emissions are coming from a surface heated to about 800 degrees.

Thomas Gold, Director of the Center for Radiophysics and Space Research at Cornell University, who enjoys championing the unorthodox, has suggested there may be oceans on Venus, protected by a heavy surface scum. There are some, as well, who are unconvinced by the Mariner II observation. They point out that our knowledge of the planet is still extremely meagre and refuse to write it off as a possible abode of life. The majority of scientists, however, now believe Venus is a desert so hot that the long-chain molecules essential to any form of life could not exist.

It has been the fashion to dismiss the possibility of life on the four big outer planets on the grounds that their air is 'poisonous' and they are too cold. Nevertheless, the atmospheres of these planets seem to resemble that which enveloped the earth during the period when life was evolving here, and our temperature readings have been obtained largely, if not entirely, from the upper portions of these atmospheres. It has been proposed that there may be a strong 'greenhouse effect' on Jupiter, in which the warmth of sunlight freely pierces the many cloud layers, but infra-red radiation from heating of the planet's surface is trapped by the clouds as though by the glass roof of a greenhouse. Since the radiation cannot escape again into space, an accumulation of warmth is postulated, despite Jupiter's great distance from the sun. Thus, according to Sagan, lower layers of the atmosphere may be above room temperature and Jupiter may have warm seas of ammonia or water. While Haldane and others have argued that a whole system of organic and inorganic chemistry could take place with liquid ammonia as the medium instead of water, the feasibility of such a system, as noted at the end of Chapter 9, has been questioned.

Jupiter, largest of the planets, has a surface gravity far stronger than that of the earth, which would make locomotion there difficult. Its powerful, irregular bursts of radio 'noise' are one of the most perplexing features of the solar system. The planet seems to have a powerful magnetic field and very intense radiation belts. In 1831 a huge red spot was observed on the planet, and since then its colour and visibility have fluctuated. The spot is some 25,000 miles long and 8,000 miles wide, and its location on the planet changes slightly as though it were adrift. Some have proposed that it is a gigantic meteorite that fell more than a century ago and is floating on a Jovian sea or within its very dense atmosphere, but this idea had been challenged.

Perhaps the most fantastic suggestion regarding life within the solar system was that of J. B. S. Haldane, who proposed, though he deemed it 'unlikely', that there could be life in the earth's interior based on the chemistry of partly molten silicates and deriving its energy from the oxidation of iron. Perhaps he was influenced by the story of Sir Arthur Conan Doyle in which a deep well was drilled, somewhat in the manner of the proposed 'Mohole'. It pierced what proved to be the earth's shell, reaching a soft, gelatinous substance inside. The earth, it turned out, was alive, like a gigantic sea urchin. In view of the story's dramatic ending, it was entitled 'When the World Screamed'.

The ideas of men like Shklovsky, Salisbury and Haldane have produced a reaction reminiscent of that when Lowell elaborated on the canals. In 1964, as though to make amends for publishing such uninhibited articles as the one by Salisbury, *Science* displayed George Gaylord Simpson's angry dissent against the American space effort – in particular its goal of learning whether or not there is life on Mars: 'I cannot share the euphoria current among so many, even among certain biologists,' he wrote, commenting wryly that some of these biologists are now 'ex-biologists converted to exobiologists'. Despite the seasonal changes and other signs of activity on Mars, he said, there is no clear evidence that life exists anywhere in the solar system except on earth.

'Wishful thinking, to which scientists are not immune,' he asserted, 'has obviously played a part here.' His comment is certainly applicable to a good deal of the speculation. The discussion of life beyond the earth is unavoidably subjective. Many hope there is life elsewhere; some get carried away by their enthusiasm, devising modes of life elsewhere that seem preposterous to the sober scientist. Nevertheless, while Simpson is right in saying we have no reliable evidence of life on Mars, neither can we say with any assurance that it does not exist.

The costly and challenging effort to learn the answer is under way. The seven-and-a-half-month flight of Mariner IV past Mars was but a curtain-raiser. The main show for the next decade will be the dispatch of a succession of Voyager spacecraft that will dwarf Mariner IV in size. It is unlikely that the first of these vehicles and the giant rockets needed to launch them will be operational before the Mars opposition of 1971, but the Russians, having shot at Mars unsuccessfully on both the last two oppositions, will probably try again for those of 1967 and 1969. A modified Mariner shot is possible for one of these oppositions and, in 1971, the United States may use its huge Saturn V rocket to launch two Voyagers together. They would be adjusted in flight to arrive ten days apart. Both might orbit Mars or one might attempt a landing.

While the very thin air of Mars will be a handicap in landing, it will be a boon in placing spacecraft in low orbits, from which they can scan the surface at close range almost indefinitely. The Jet Propulsion Laboratory of the California Institute of Technology, which carried out the Mariner project so brilliantly on behalf of NASA, cannot take any chances that one of the orbiters will encourage drag from the upper air of Mars and drop out of orbit, contaminating the surface. The orbiters must be kept well above this danger zone. Not only can Mars be mapped from orbit, but much can be learned about its surface and atmosphere in this manner.

Nevertheless, the argument as to whether or not there is life on Mars will not be settled until its surface has been closely examined. The immediate goal is the development of 'Automated Biological Laboratories' to be landed intact, using such far-fetched devices as enormous balloons to cushion the landing. At the same time ingenious systems have been developed to test the surface for signs of life. Such detectors, it is universally agreed, must be

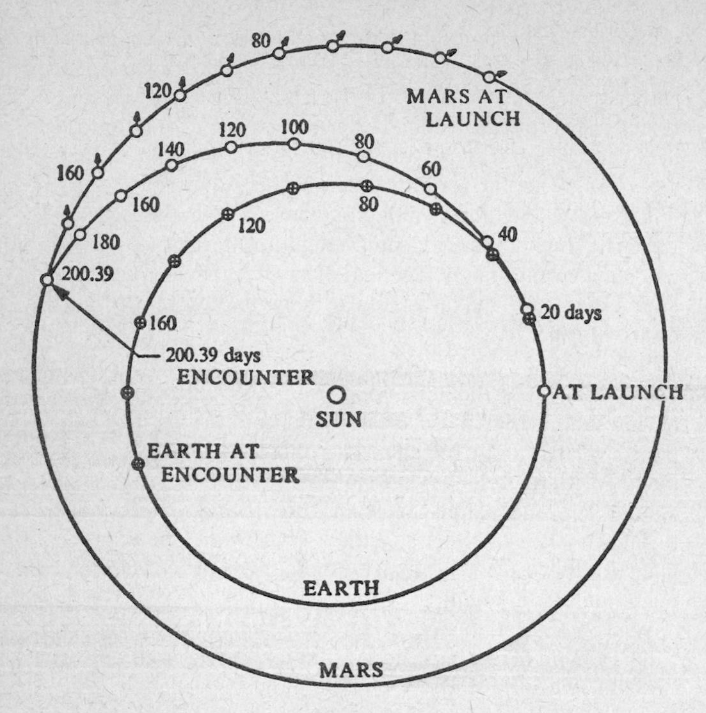

A typical Earth-Mars vehicle trajectory with a 200-day travel-time. The figures along each path denote the number of days after the launching. The earth is designated by a cross enclosed in a circle. Mars is marked, at each position, by an arrow symbol.

aboard the first vehicle to land, for no matter how carefully it is sterilized, it might carry a microbial stowaway. By the next landing such an organism could have spread widely.

The detectors, in general, are designed to look for the functioning of life, rather than life forms. As one participant in the programme has put it: 'You don't look at the thing itself, you look at what it does.' The detectors depend on metabolic processes typical of life on earth, but the most diverse processes have been chosen, including those performed in the absence of oxygen. The following are the three most talked-about devices:

Gulliver: This instrument, developed by Norman H. Horowitz, Professor of Biology at the California Institute of Technology, and Gilbert V. Levin, contains three baby cannon that shoot 25-foot strings out across the landscape, much like the line-throwing guns on ships. The strings are sticky, so that, as they are reeled back into the detector, they will pick up particles, including perhaps microbes. The strings, when fully reeled in, are soaked in a nutrient broth.

To see if any of the broth is 'eaten', its carbon compounds are tagged with radioactive carbon in the same way as those used by Calvin in his experiments on photosynthesis and the origin of organic molecules. If Martian bacteria feed on the broth, they will release carbon dioxide that, in this case, is radioactive be-case of the tagging. A Geiger counter measures the radioactivity of gases produced by the process and, via a radio system, reports whether there is any lifelike activity.

Multivator: This unit has been designed by a group at Stanford University Medical Center led by Joshua Lederberg. Like Gulliver, it weighs only about a pound. Soil samples are blown into various chambers. Certain chambers contain substances that are acted upon by enzymes characteristic of life as we know it (in particular the enzyme phosphatase, which breaks down the phosphates). The chemicals in the chamber will fluoresce if the enzyme is present on Mars. The Multivator has fifteen experimental compartments, allowing for a variety of experiments as well as 'control' chambers to check against contamination or damage to the system en route.

Wolf Trap: This name was inevitably given to the detector devised by Wolf Vishniac, Professor of Biology at the University of Rochester. Soil or dust is sucked into its test chamber, which is then closed and a culture broth introduced. The presence of microscopic life can be detected either by observing increased cloudiness of the broth (as in the experiments of Pasteur's day) or by changes in its acidity.

A variety of other experiments have been considered, including television transmission of microscopic views of Martian material and systems to detect optical activity – the twisting of light waves described in the previous chapter. Such systems are probably too complex and unreliable to go on the first payloads. Nevertheless, at the Warsaw meeting Horowitz pointed out the importance of learning whether the chemicals of life on Mars – if there are any – rotate light in the same way as their counterparts on earth. The chemistry of life on our planet, he said, is 'remarkable for its monotony' in that every species, from virus to man, employs similar proteins and nucleic acids. As noted earlier, the amino acids forming our proteins always twist light to the left; the sugars in our nucleic acids always twist them to the right. Thus in a fundamental sense, said Horowitz, 'there is only one form of life on earth'. This uniformity may, in fact, mean that all life on this planet is descended from a single ancestor. Life elsewhere could be the mirror image of our own, with chemicals that rotate light in the opposite directions. To discover such life on Mars would be a sure indication that it has arisen independently.

Martian life may also differ in other ways, including an absence of sex. As pointed out in a recent Federal study, it may be that life on Mars has found some means, other than sexual reproduction, to achieve genetic variations, and hence evolution. This study also produced one of the more novel proposals for remote exploration of Mars, namely by laser. It was noted that these newly invented light generators might soon be able to produce a beam of light on a single wavelength so narrow and intense that it could be used to scan small portions of the planet, providing new information on the chemical composition of its surface.

Another possibility, discussed by the developers of Gulliver, is that life on Mars, instead of being based on the bag of liquids that we call a cell, may be in a solid state, like the 'brains' of our computers. If living things there stand as motionless as statues, how are we to recognize them?

One scheme for the detection of Martian life, be the latter mobile or stationary, is that of Gold at Cornell. He argues that, if a planet like the earth were lifeless, there would be few vertical surfaces. That is, it would lack trees, plants, animals, people, buildings, etc. If one viewed the earth in its equatorial plane from a great distance, some of the vertical surfaces on the western rim would be approaching at the rotational speed of the earth, which in low latitudes is more than 1,000 miles an hour. The surfaces on the opposite edge would be receding at the same speed. These relative motions would produce frequency shifts in radar echoes and the strength of these altered, or 'Doppler-shifted', echoes would indicate the richness of the planet in vertical surfaces. By tuning the wavelength until the effect reached its maximum it might, according to Gold, be possible to estimate the height of things that grow on another planet.

It has also been argued that, in concentrating on detection of microscopic life, we may overlook large creatures that are moving about – 'Martian elephants'. Large life forms are unlikely to be present if microscopic life is absent, whereas, as noted earlier, it is thought more likely that life on Mars has not evolved past the miniature level. Nevertheless it has been proposed that microphones be included on early payloads to listen for 'the rustle of leaves' or, possibly, 'squeaks', and television scanners to look for such signs of advanced life as footprints and 'minarets'.

It may well be that the question of whether there is life on Mars will not be settled until biologists themselves walk its surface. NASA has contracted with various groups to draft plans for hypothetical expeditions. For example, Douglas Aircraft has been given $91,901 to study a mission to take place during the decade after 1975 (that is, following the Voyager shots). From three to ten astronauts would make the trip, some of them descending to the surface from an orbiting space ship to remain for from ten to fifty days. The round trip would take from one to three years.

Many such voyages may be necessary to explore the planet adequately, for, with no oceans, its land surface is three times as large as North and South America combined. The exploration will be expensive and probably would be an intolerable burden on our budget until military expenditures are drastically reduced.

The inevitability of manned exploration of Mars was stressed at a historic meeting, organized in the summer of 1962 by the Space Science Board of the National Academy of Sciences. For eight weeks, at the State University of Iowa, more than 100 specialists in a broad range of fields discussed the goals of American space research. Part of the report was prepared by the conference's Working Group on Biology under the chairmanship of Allen H. Brown, Professor of Botany at the University of Minnesota. It described the search for extraterrestrial life as 'the prime goal of space biology'. While sober care is necessary in planning the space programme, the report said, this should not stifle 'that element of challenge and high adventure that has characterized the scientific enterprise since it was put on its modern course by the brave minds of the Renaissance'.

The search for life beyond the earth must be directed at Mars, the report added, for if we find life there and show that it originated locally, we will be far better able to assess how likely is its appearance on planets in other parts of the universe. 'Arising twice in a single planetary system it must surely occur abundantly elsewhere in the staggering number of comparable planetary systems.'

In response to grumbling at the prospective cost of Mars exploration, both in dollars and talent, the National Academy of Sciences, in 1964, asked a panel of thirty-six specialists to assess the merits of such an effort. The resulting report described our present biology as provincial in that it deals with life on a single planet. Does the chemistry of life always evolve in the same way? Are its proteins invariably composed of the same twenty amino acids? Is the DNA molecule the only way that nature has achieved the continuity of life?

'To the extent that we cannot answer these questions,' said the report, 'we lack a true theoretical biology, as against an elaborate natural history of life on this planet.' Perhaps chemical evolution on Mars was blocked before it produced life, but by deciphering it we may learn to what extent our own history has been typical.

With heady enthusiasm those at the 1962 conference turned their backs on the sceptics. The search for extraterrestrial life, they said, 'is, in the opinion of many, the most exciting, challenging, and profound issue, not only of this century but of the whole naturalistic movement that has characterized the history of western thought for 300 years. What is at stake is the chance to gain a new perspective on man's place in nature, a new level of discussion on the meaning and nature of life.' If planets are common throughout the universe, they added, life may be almost as common. 'It is surely unnecessary to belabour the immensity of this prospect for any man's philosophical position, or to belabour the pusillanimous and provincial viewpoint that would shrink from pursuing it.'

13

The Uniquely Rational Way

ON 19 September 1959 the British journal *Nature* published a proposal that startled the scientific world. The authors, two physicists of impeccable reputation, suggested that at this very moment intelligent beings in some distant solar system might be trying to communicate with us. Furthermore, they presented an argument pointing to one precise radio frequency as the uniquely logical channel for such communication. They proposed that radio telescopes be aimed at certain nearby stars in search of the hypothetical signals.

Actually the discussion of possible methods for communication with other worlds long predates the discovery of radio waves. During the many decades when even the most conservative scientists believed intelligent life might exist on Mars – or even the moon – there was talk of ways to reveal the presence of intelligence on earth. In the last century the mathematician Karl Friedrich Gauss is said to have proposed that broad lanes of forest be planted in Siberia, forming a gigantic right-angled triangle. The inside of the triangle would be planted in wheat to give it a uniform colour. A modification of this plan was to erect squares on each side of the triangle, forming the classic illustration of the Pythagorean theorem.

In Vienna the astronomer Joseph Johann von Littrow is reported to have suggested that canals be dug in the Sahara, forming geometric figures 20 miles on a side. At night kerosene was to be spread over the water and set on fire. In France Charles Cros urged that the government construct a vast mirror to reflect sunlight towards Mars. A related scheme was for a system of mirrors that could be rearranged, periodically, in a form of slow-speed semaphore.

The discovery of radio waves offered a far more realistic means of interplanetary communication, and this possibility excited two of the men who pioneered in this field, Marconi and Tesla, to the point where both suspected that they had heard signals from another world. Although far less well known than Edison, Nikola Tesla was an inventive genius whose contributions to the application of electrical energy were of major importance. He seems, for example, to have been the first to propose an effective way to use an alternating current – a scheme strongly opposed by Edison, but one that made possible the harnessing of the power of Niagara Falls. Tesla invented a multitude of dynamos, motors, transformers and the like. He performed some of the earliest experiments in radio communications, apparently being the first to use transmitting and receiving antennas tuned to the same frequency.

He was also given to grandiose schemes and strange ideas. In 1917, for example, he said: 'We will deprive the ocean of its terrors by illuminating the sky, thus avoiding collisions at sea.' One of his grandest ideas was to set the electric field of the entire planet aquiver. He believed the world had an electric charge and that he might be able to generate oscillations within it. To this end, and with financial aid from John Pierpont Morgan, he set up a laboratory in Colorado Springs, Colorado, in 1899, equipped with a 200-foot transmission tower and high-voltage equipment. One night, when he was alone in the laboratory, Tesla observed 'electrical actions' that, he later reported, appeared to be signals:

> The changes I noted [Tesla said] were taking place periodically, and with such a clear suggestion of number and order that they were not traceable to any cause then known to me. I was familiar, of course, with such electrical disturbances as are produced by the sun, Aurora Borealis and earth currents, and I was as sure as I could be of any fact that these variations were due to none of these causes. . . . It was some time afterwards when the thought flashed upon my mind that the disturbances I had observed might be due to intelligent control. . . . The feeling is constantly growing on me that I had been the first to hear the greeting of one planet to another.

When Tesla was asked by the American Red Cross, at the turn of the century, to predict likely developments of the next hundred years, he replied on 7 January 1900 with a reference to these apparent signals:

Faint and uncertain though they were [he wrote], they have given me a deep conviction and foreknowledge, that ere long all human beings on this globe, as one, will turn their eyes to the firmament above, with feelings of love and reverence, thrilled by the glad news: 'Brethren! We have a message from another world, unknown and remote. It reads: one . . . two . . . three . . .'

A few decades remain before the century ends and we can assess the accuracy of this prediction. It is noteworthy that the 1959 proposal for a listening programme envisaged that intelligent beings in another world would use just such a device for attracting attention – a series of pulses representing such numbers as one, two, three, etc. Had it not been for Tesla's personality, his report might have made more of an impression. He was an eccentric figure who allegedly believed in mental telepathy and displayed a strange affinity for pigeons, in particular a certain white bird that, in his later years, he spoke of with peculiar passion. He implied that he had had a closer rapport with this pigeon than with almost any human being and that, when it died, he could no longer work creatively. He fed the pigeons in Bryant Park, in the heart of Manhattan, with devoted regularity and, toward the end, when too weak to go himself, he paid a Western Union messenger to do so.

Tesla was seized upon by the spiritualists, although he does not seem to have been one himself, and was the hero of a book published in 1959 under the title *Return of the Dove* (his pigeon thereby being transformed into a somewhat different species). The book, by Margaret Storm, described him as having been born aboard a space ship en route from Venus to the earth in 1856. He was deposited, it said, in a remote mountain province of what is now Yugoslavia. Tesla was, in fact, a Croatian by birth.

While Tesla was experimenting in Colorado, the Italian physicist Guglielmo Marconi demonstrated the capabilities of radio by sending messages between England and the Continent. By 1901 his radio waves had spanned the Atlantic and a revolution in communications was under way. During World War I transatlantic radio signalling was routine and in 1920 Marconi reported that the stations of his company, on both sides of the Atlantic, had been hearing strange signals from time to time since before the war. Some were in what sounded like a code, but were meaningless. Most frequently heard was the letter V (three dots and a dash). The latter is frequently used by radio operators on earth as a test signal for tuning their equipment. Marconi was asked by a reporter if these might be signals from another planet, and he was quoted as saying Yes. On 2 September 1921 the front page of the *New York Times* carried a report from one of Marconi's associates that the physicist was convinced some of the signals came from Mars. He pointed out that their frequency was ten times lower than that of man-made signals then in use. Tesla commented that such signals could be 'undertones' of man-made signals, comparable to the harmonics of a musical note. Others said all of the interference with transoceanic communications could be accounted for by the distant effects of sunspots. There was a dearth of further comment from Marconi himself, but these events built up great excitement at the coming close opposition of Mars in 1924. The Sperry Gyroscope Company, which had become eminent in its field during World War I, proposed that its new, high-powered searchlights could be clustered to generate a beam powerful enough to be visible on Mars.

In anticipation of the close passage of Mars, David Todd, who had been head of the Astronomy Department at Amherst College, in Massachusetts, proposed, with the aid of a wealthy collaborator, B. McAfee, to convert an abandoned mine shaft in Chile into a mammoth telescope. The shaft, almost 60 feet in diameter, was so located that Mars would pass directly over its mouth. As a mirror at the bottom of the shaft, Todd proposed to spin a basin of mercury 50 feet in diameter fast enough so that the mercury would assume the proper concavity. The telescope, he claimed, would bring the surface of Mars, which was to pass at a distance of 34,630,000 miles, to within less than 2 miles of his instrument. Astronomers at Harvard and Yale denounced the plan as 'preposterous' and 'a foolish, wild scheme'. They pointed out that such magnification would only produce a meaningless blur.

Todd's next proposal was to persuade all radio stations on earth to shut down, during the close passage, and listen for signals from the other planet. The possibility of a civilization on

Mars superior to our own was still sufficiently in men's minds for Todd to have some success. On 21 August 1924 the Chief of Naval Operations of the United States Navy sent a dispatch to the twenty most powerful stations under his command, from Cavite in the Philippines to Alaska, the Canal Zone and Puerto Rico, telling them to avoid unnecessary transmissions and to listen for unusual signals. A similar order was sent to the Army stations and the Executive Officer of the Army Signal Corps announced that William F. Friedman, Chief of the Code Section in the Office of the Chief Signal Officer, was standing by to decipher any messages received. He expressed confidence that Friedman could do so. Elsewhere a man who had been experimenting with radio transmission of photographs tuned up his equipment in case Mars wanted to send us pictures. There was, of course, considerable scepticism in the government and in other circles. Only one or two commercial broadcasting stations seem to have closed down for the proposed periods of five minutes each hour during the passage. The more conventional astronomers were aghast at the whole thing. The *New York Times* predicted editorially that if a great many listeners tried to pick up Martian signals, some were sure to hear them on the basis of the old adage that the wish is father to the thought. Indeed there were such reports. Some from Vancouver, British Columbia, proved later to have come from new United States radio beacons. 'Harsh dots' were reported by one of the most powerful receivers in Britain, but were quickly dismissed as being of natural origin. The bubble of excitement collapsed and it was not until past mid-century that the possibility of communicating with other worlds took on a new dimension and new respectability. Two major developments were responsible for this: the birth of radio astronomy and the explosive evolution of radar.

It was in the early 1930s that it was first proposed that emissions at the extreme short-wave, or 'microwave', end of the radio spectrum, where it approaches the even shorter wavelengths of light, might be focused into an invisible, penetrating beam that could search out enemy targets. In 1935 the British Air Ministry authorized the secret construction of five stations along the east coast of England, designed to sweep microwave beams across the sky to warn of any surprise attack by Germany's new *Luftwaffe*. Two years later another fifteen of these stations were installed and, during the war that followed, this equipment, which came to be known as radar, proved so valuable that there was a massive research effort in electronics. Transmitters, receivers and antennas of ever greater efficiency were developed until, by 1946, it was possible to bounce radar signals off the moon. By 1959 ten groups in six countries had studied the moon with radar echoes, observing, for example, its surface properties in a manner impossible through optical telescopes.

The use of radar for such astronomical work in the United States was a by-product of the research, largely financed by the Air Force, which sought to develop equipment that could give the country as much warning as possible of a missile attack. One of the most powerful instruments in this programme was the parabolic, or dish-shaped, antenna, 84 feet in diameter, mounted on a tower atop Millstone Hill, near Westford, Massachusetts. It was operated by the Lincoln Laboratory of the Massachusetts Institute of Technology, which, in February 1958, used it for the first time in an effort to obtain radar echoes from another planet. It was aimed at Venus, our nearest neighbour beyond the moon, and prolonged analysis of the tape-recorded results seemed to show an echo. President Eisenhower sent a congratulatory message and, although the analysis is now thought to have been erroneous, the experiment awakened scientists to the growing power of the world's transmitters. When Venus came close to the earth again in 1959 new attempts were made, both by Millstone Hill and by the world's largest antenna at Jodrell Bank in England. Again the results were unconvincing, but when Venus made its close approach in March and April of 1961 a number of antennas in England, the Soviet Union and the United States, including an 85-foot dish at Goldstone, California, obtained echoes.

Two such antennas at Goldstone, set seven miles apart in the Mojave Desert, constitute the central unit of the Deep Space Instrumentation Facility, operated for NASA by the Jet Propulsion Laboratory of the California Institute of Technology. The station is one of three located around the world to keep continuous track of American space vehicles. After several weeks of preparation and testing, the equipment was ready on 10 March and the two antennas were aimed at Venus, one to transmit and the other to receive. Some six and a half minutes after the transmitter was turned on, the stylus recording incoming noise on a moving roll of paper moved over slightly – indicating the arrival of a weak echo.

Both antennas were controlled by gears that kept them aimed squarely at Venus, despite the planet's motion across the sky. After a half hour, to find out whether or not the incoming echoes were genuine, the motion of the transmitting antenna was halted and Venus slowly moved out of its beam. Six and a half minutes later the stylus moved back to the intensity level it had shown before the experiment. The interval was that required for radar waves to travel to Venus and bounce back.

The transmissions were at 2,388 megacycles per second with a power of 12,600 watts. About 10 watts of this total hit the surface of Venus. It was calculated that 9 watts were absorbed and 1 reflected back into space. Of this, one-hundredth of a billionth of a billionth of a watt returned to strike the receiving antenna, yet this was still 10 times the background noise, making it observable.

Within weeks other observatories, including Millstone Hill, were obtaining echoes, and the latter was able to re-examine earlier records and discover, hidden there, true echoes from the Venusian surface.

Thus did earth's inhabitants send their first signals to another planet and receive 'answers'. Among the factors which made this possible was the use of radio frequencies in which the sky is comparatively quiet. Another was the availability of two revolutionary new devices for amplification of returning echoes. If you turn up the volume of an ordinary radio, the hum generated by its amplifiers becomes so loud that it drowns out any weak signals that might otherwise be received. The new devices, known as the maser and the parametric amplifier, eliminate almost all of this receiver noise. Both were used with the Goldstone antenna.

Meanwhile radar echoes had been obtained from the sun and other bodies of the solar system, but still more dramatic evidence

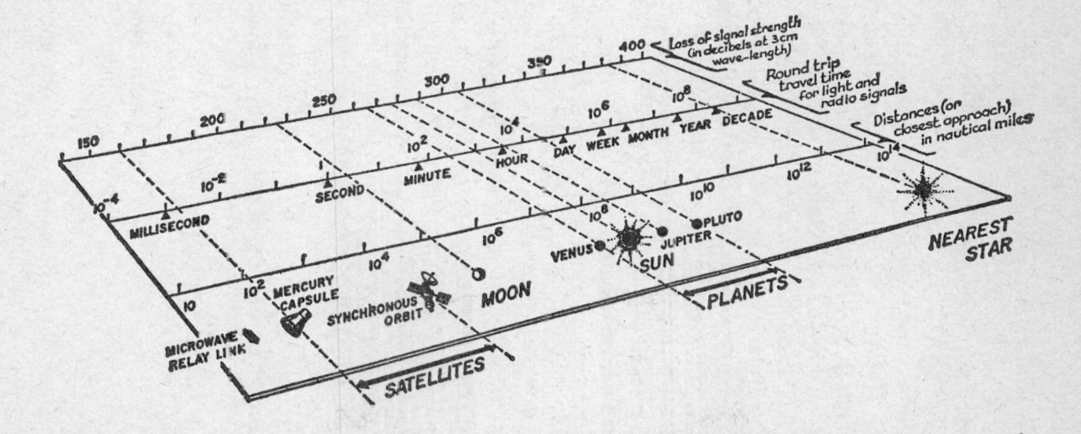

This diagram, laid out like a football field in space, is based on one devised by M.I.T. to illustrate the power loss, time lag and vast distances involved in the use of radar and radio on an astronomical scale.

of the capabilities of radio for long-range communications came from radio astronomy. It was shown that radio waves traverse the vast reaches of the universe even more readily than light waves, and eventually this new science pointed to what seemed the uniquely rational wavelength on which to search for signals.

Radio astronomy came into being by accident through the efforts of Karl G. Jansky, of the Bell Telephone Laboratories, to track down the high-frequency static that was interfering with his company's transoceanic communications. At a field station in Holmdel, New Jersey, he built an antenna array on a wooden frame 100 feet long which rode on four wheels salvaged from a Model T Ford. A motor rotated this array, which he called his Merry-go-Round, once every 20 minutes. He began his observations in August 1931 on a wavelength of 14·6 meters (20,600 kilocycles) and soon tracked down the sources of two types of static. One obviously came from the lightning in nearby thunderstorms. The second he attributed to distant storms whose radio emissions were probably bent back to earth by the ionized regions of the upper air. A third form was quite different, producing a loudspeaker hiss whose intensity slowly changed during the day. He reported in 1932, in the *Proceedings of the Institute of Radio Engineers*, that the direction from which this hiss came travelled 'almost completely around the compass in twenty-four hours'. In the previous December and early January, he said, its direction coincided generally with that of the sun, his array not being able to pinpoint the source very precisely. Then he noticed that its direction was shifting and, as of his writing, on 1 March he said, 'it precedes in time the direction of the sun by as much as an hour'.

Nevertheless he felt the source was most probably some effect of sunlight on the earth's atmosphere. The shift in direction might be related, he said, to the seasonal rise of the sun to higher elevations in the sky. However, an astronomically-minded colleague suggested that Jansky rework his data, using celestial coordinates (the 'latitude and longitude' of the stars on the heavenly sphere) to designate the apparent direction of the source each day. He made the exciting discovery that it remained fixed among the stars.

At first there seemed to be a jump in his record until he realized that he had forgotten to allow for the shift from Standard to Daylight Saving Time.

At a meeting of the American Section of the International Scientific Radio Union on 27 April 1933 he announced his conclusion. The radio emissions, he said, appeared to be coming from beyond the solar system. They might be from a single source, he added, or 'from a great many sources scattered throughout the heavens', the direction at which he observed maximum intensity being simply the centre of this activity. He noted that this direction, toward the constellation Sagittarius, the Archer, was that of the centre of our galaxy. It is now thought that both of his suggestions are correct. The direction that he identified is that from which intense radio emissions indicate some incredibly fierce, but as yet unknown, activity in the core of the galaxy – it is known as Sagittarius-A. But it was also soon shown that radio emissions are coming from the entire galaxy, as well as from the most distant objects in the universe of which we have knowledge – objects whose nature is being hotly debated, for they emit radio waves of such intensity that physicists find it difficult to account for their generation.

Jansky's report of signals from beyond the solar system was front-page news in the press, even though he discounted the possibility that they were of artificial origin. The astronomers, apart from two young men at Harvard (Jesse L. Greenstein and Fred L. Whipple), gave it the cold shoulder. It was for an amateur astronomer and radio 'ham', Grote Reber, to make the next big step. Reber was, by profession, a radio engineer and, at his own expense, he built a parabolic antenna in his backyard at Wheaton, a Chicago suburb. Being 31 feet in diameter, the dish crowded his yard. It was the first such antenna turned on the heavens and his results, published in 1940 and thereafter, showed virtually the entire Milky Way to be a source of radio 'noise', with several areas of intense emission.

World War II, which impeded further progress in radio astronomy, also gave it a new stimulus. In February 1942 British Army radar operators complained of a new form of German jamming. The Germans had been trying to nullify the British warning system by flooding the radar receivers with signals on their echoing frequencies. The new interference was referred to J. S. Hey of the Army Operational Research Group, who noticed that the direction from which the jamming came did not point to one or two transmitters on the coast across the Channel. Instead, all the directions pointed to a single source: the sun. It happened that there was at this time a very large sunspot, and Hey guessed that it was the villain, not the Germans. In that same year solar emissions were also identified in the United States, but anything to do with such high-frequency research was considered a military secret, so radio observations of the sun did not come into their own until after the war.

Meanwhile, in 1944, word of Reber's observations had reached the German-occupied Netherlands and had excited the interest of astronomers at the Leiden Observatory. Unlike American scientists, who were preoccupied with the war effort, the Dutchmen were free to dabble in pure science and Jan H. Oort, director of the observatory, held a seminar on the implications of Reber's observations. Since radio waves are an extension of light waves into longer wavelengths, the discovery that parts of the sky 'shine' in the radio portion of the spectrum meant that an entirely new window had been opened on the heavens. Until then all of our knowledge of the universe, apart from a few clues culled from meteorites or cosmic-ray particles, had been acquired through observation of that narrow band of the electromagnetic spectrum which we call light.

However, Oort pointed out, one drawback to the radio spectrum seemed to be that, unlike light, it lacked the sharp emission (or absorption) lines that had proved so useful in astonomy. These lines, the by-products of various atomic processes, could be used, as noted earlier, to detect relative motion toward and away from the observer and many other phenomena.

As though picking up this challenge, one of Oort's students, Hendrick Christoffel van de Hulst, came up with the proposal that clouds of individual hydrogen atoms, as opposed to the paired atoms of hydrogen gas, should emit radio waves at a wavelength of 21 centimetres. Since it was suspected that such clouds exist widely in space, there should be a sharp augmentation of cosmic radio noise at that wavelength – in other words, an emission line.

The hydrogen atom consists of an electron and a proton, both in effect spinning and thus acting like tiny bar magnets. Because like repels like in neighbouring magnets, the most natural alignment of these particles is with their magnetic poles pointed in opposite directions. It therefore takes a bit of energy to flip the electron over so that its positive pole is aligned with the positive pole of the proton. When such a flip has occurred, the atom has a slight reserve of energy. Ultimately the electron flips back, emitting this energy as a radio wave. The waves oscillates at a characteristic frequency of 1,420,405,752 times a second (1,420 megacycles), which corresponds to a wavelength of 21 centimetres. In the lonely reaches of space, where hydrogen atoms are free from strong energy inputs and magnetic influences, they still receive nudges of energy from collisions and radiation. Van de Hulst proposed that hydrogen clouds throughout the universe must be shedding this energy as radio waves at 21 centimetres. By good fortune, this wavelength passes freely through space and the earth's upper atmosphere. Longer waves are apt to be absorbed in the upper air.

The next step was to look for such emissions. For a number of years Edward M. Purcell at Harvard had been studying the radio-frequency resonances of atomic nuclei. In 1945 he devised a technique that made it possible to measure the utterly tiny magnetic fields generated by the spin of these nuclei, using a high-frequency coil that could be tuned to match the resonance frequency of the nucleus under study. For this he and Felix Bloch, who had done similar work at Stanford University, shared the Nobel Prize in Physics for 1952.

It was in the previous year that Purcell, with his Harvard colleague Harold I. Ewen, sought to detect the 21-centimetre hydrogen emissions predicted by van de Hulst. At a cost of $400 a university carpenter built a horn antenna on the laboratory roof. The electronic equipment, said Purcell, 'was all scrounged', and on 25 March 1951, when they turned on the receiver, the predicted emissions were there. Less than two months later the Netherlands astronomers obtained similar results and both groups found that the emissions, instead of showing up as a single, narrow line at 21 centimetres, were spread out in wavelength, with several peaks of intensity.

What they were seeing were hydrogen clouds moving at different velocities toward and away from the receiving antennas, with consequent modification, or 'Doppler-shifting', of the wavelengths. These relative motions were produced by the earth's spin, by our planet's orbital flight around the sun, by the sun's flight around the core of the galaxy and by inherent movements of the hydrogen clouds themselves. The relative contributions of all these factors could be determined and distances to the various clouds could be indirectly estimated. Soon the Dutchmen were mapping spiral arms of our galaxy, hidden from us by dust clouds but revealed by their 21-centimetre emissions. The dust that closed much of the universe to our telescopes had suddenly become transparent.

The discovery of the 21-centimetre emissions gave a tremendous boost to radio astronomy. Large antennas were already under construction or planned, for radio waves from space had opened up many lines of investigation. They signalled eruptions on the sun, even in cloudy weather; they indicated the surface temperatures of the moon and nearby planets; they disclosed the existence of atomic particles trapped and gyrating furiously within distant magnetic fields, as they do in the radiation belts surrounding the earth and Jupiter or in the turbulent gas clouds of the Crab Nebula.

In England the mammoth radio telescope at Jodrell Bank was already under construction when word was received that the 21-centimetre emissions had been observed. The diameter of the British dish is 250 feet (a football field is 300 feet long), and it is mounted on a frame that can be tilted and rotated to aim the antenna at any point in the sky. To achieve rotation, the array, weighing 2,000 tons, rides a circular, double railroad track. To tilt the 800-ton dish assembly the British installed a drive taken from one of the main battery gun turrets of the battleship *Royal Sovereign*. Protruding 62½ feet from the apex of the dish is a mast that holds the receiving equipment at the focal point. The parabolic shape of the dish, like that of the mirror in an optical telescope, reflects all radio emissions from the direction of its aim to this focal point. The dish was designed for wavelengths so long that a coarse wire mesh was sufficient to reflect them to the focus. The plans were hastily revised to provide a surface of solid metal sheets that would reflect waves as short as 21 centimetres.

In the United States the only large dish capable of observing at that wavelength was the one, 50 feet in diameter, on the roof of the Naval Research Laboratory, across the Potomac River from Washington's National Airport. Hence it was decided that the first big antenna at the projected National Radio Astronomy Observatory in West Virginia should be designed for wavelengths of 21 centimetres and even less.

This is how matters stood when the two physicists began formulating their proposal that there be a search for intelligent signals from another world. They were Giuseppe Cocconi and Philip Morrison, both of them professors on the faculty of Cornell University in Ithaca, New York. In the spring of 1959 Cocconi was working on a paper that he was to present at an international conference on cosmic rays, to be held that summer in Moscow. It was his thesis that some celestial objects, such as the Crab Nebula, should be strong emitters of gamma rays as an extreme form of what is known as synchrotron radiation. If one should be so unwise as to peek into one of the very high-energy circular electron accelerators, such as the synchrotron at the California Institute of Technology that develops 1·1 billion electron volts, one would see a strange and brilliant light. This is synchrotron radiation and represents energy given off by the electrons because they are forced, by magnets, to fly in a circle. The radiation is given off throughout much of the electromagnetic spectrum, including visible light, radio waves and gamma rays.

Both the light and the radio emissions from the Crab Nebula show that they are products of synchrotron radiation. This nebula is one of the most awesome sights of the sky, for in photographs it looks like a snapshot of an explosion. That, in effect, is just what it is, for the Crab Nebula is the remnant of a supernova first seen in 1054, when it was visible by day as well as by night. The gases and other debris are still flying outward at great velocity, and within these clouds are magnetic fields that have trapped gyrating electrons, producing the radiation.

One of Cocconi's points was that the gamma rays from this nebula should stand out boldly against the sky because such emissions must be a rarity in the heavens. He proposed a gamma-ray detector that could, in a rough way, determine the direction from which such rays were coming.

One evening, as he discussed this with his wife, Vanna, herself a physicist at Cornell, it occurred to them that the scarcity of gamma-ray emitters in the sky pointed to these rays as a promising vehicle for interstellar signalling. If an unexpected source of gamma rays were observed, would it not be worth examining very closely?

The Cocconis had often discussed with Morrison the possibility of intelligent life in other worlds. Morrison had been a group leader at Los Alamos during the effort there to produce the first atomic bomb. He jumped eagerly at the suggestion that it might be a good idea to look for signals, rather than try sending any. However, he argued against gamma rays as difficult to generate and receive. A synchrotron adequate to generate a beam of gamma rays would probably cost millions of dollars (no such device has yet been built), whereas radio transmitters are so cheap that taxis all over the world carry them.

The radio frequencies that travel readily through space and through the atmospheres of planets like our own lie in the range between 1 and 10,000 megacycles. The most powerful instrument for receiving signals in this general region was that at Jodrell Bank, now in operation more than a year. From CERN, the international atomic research centre in Switzerland, Cocconi wrote to Sir Bernard Lovell, founder and director of the great radio astronomy observatory, with a proposal that Sir Bernard at first dismissed as 'frivolous'. The letter, dated 29 June 1959, began as follows:

Dear Dr Lovell,

My name is probably unknown to you, so let me start by saying that I am now at CERN for one year, on leave from Cornell University, where I am professor of Physics.

Some weeks ago, while discussing with colleagues at Cornell the emission of synchrotron radiation by astronomical objects, I realized that the Jodrell Bank radio telescope could be used for a programme that could be serious enough to deserve your consideration, though at first sight it looks like science fiction.

It will be better if I itemize the arguments.

(1) Life on planets seems not to be a very rare phenomenon. Out of ten solar planets one is full of life and Mars could have some. The solar system is not peculiar; other stars with similar characteristics are expected to have an equivalent number of planets. There is a good chance that among the, say, 100 stars closest to the sun, some have planets bearing life well advanced in evolution.

(2) The chances are then good that in some of these planets animals exist evolved much farther than men. A civilization only a few hundred years more advanced than ours would have technical possibilities by far greater than those available now to us.

(3) Assume that an advanced civilization exists in some of these planets, i.e., within some 10 light years from us. The problem is: how to establish a communication?

As far as we know the only possibility seems to be the use of electromagnetic waves, which can cross the magnetized plasmas filling the interstellar spaces without being distorted.

So I will assume that 'beings' on these planets are already sending towards the stars closest to them beams of electromagnetic waves modulated in a rational way, e.g. in trains corresponding to the prime numbers, hoping for a sign of life.

Cocconi then went on to point out that planets in orbit around even the nearer stars would still appear so close to them, as seen from the earth, that signals originating from such planets would be lost in emissions from the star itself unless the wavelength were in a part of the spectrum where the star was comparatively quiet. This, he argued, limits the suitable wavelengths to those in the radio portion of the spectrum and to the short-waved end of the gamma rays. Of these the radio waves seemed the logical choice because they are so much easier to send and receive.

The transmitted signals, he pointed out, must be strong enough to stand out clearly against the background emissions of the sky as a whole and must also be powerful enough to survive their very long journey to the solar system. Their loss of strength, in this respect, is determined by the so-called inverse square law, which applies to any form of electromagnetic radiation, be it light, radio waves or gamma rays. The law says that the intensity falls off according to the square of the distance from the source. Thus if you increase your distance from a light fourfold, it becomes sixteen times dimmer (see the drawing on p. 203).

In view of these considerations, said Cocconi, if a radio dish the size of that at Jodrell Bank were used to transmit from a distant planet and the British instrument were on the receiving end, the power required to transmit detectable signals would lie beyond the reach of our present technology. 'But I want to have faith,' he wrote, 'and will assume that they have larger mirrors and more powerful emitters and can do it.' What he therefore hoped Lovell would do, he said, was 'a systematic survey of the stars closest to us and spectroscopically similar to the sun, looking for man-made signals'.

'As I said before,' he wrote in conclusion, 'all this is most probably fiction, but it would be most interesting if it were not.'

'I leave to you the judgement on the feasibility of such a search.'

Lovell's reply on 14 July was brief and, in Cocconi's words, 'rather disappointing'. Lovell said such a search would be 'difficult', and added that among current tasks of the radio telescope was 'a survey of certain flare and magnetic stars'. When that was done, he said, 'perhaps we shall have time to look at a few others.' As the years passed, Lovell warmed to the idea and even published Cocconi's letter as an appendix to one of his books. In the *New York Times Magazine* of 24 December 1961 he wrote that he 'would still find it difficult to justify the diversion of any of the world's present radio telescopes to such speculative work.' Nevertheless, he continued, 'during the past two years or so the discussion of the general problem of the existence of extraterrestrial life appears to have become both respectable and important.'

The cool reception that Lovell intitially gave to the proposal did not discourage Morrison and Cocconi. Both men went to the cosmic-ray meeting in Moscow, and by this time Morrison had an idea that opened the way for a far more precise suggestion. He had discussed the problem with colleagues at Harvard, where excitement at discovery of the 21-centimetre line in the radio spectrum was still running high. The weakness of Cocconi's original proposal was the vastness of the spectrum that had to be searched. If you spin your radio dial, a powerful broad-band signal is easy to spot, but the transmitted energy is spread over a wide range of frequencies. The narrower the band, the farther away is a signal audible above the background noise, assuming that the receiver is correctly tuned. But the narrower the band the more difficult it becomes to find the signal, particularly if you do not know what frequency to look for.

Searching the entire range of possible frequencies, as Morrison said later, was like trying to meet a friend in New York without having a prearranged rendezvous. It would be absurd to wander the streets in the hope of a chance encounter. The sensible approach was to put yourself in your friend's place and guess what logical meeting place he might select, such as the information booth in Grand Central Terminal. What had to be found, Morrison said, was the 'uniquely rational way' for interstellar attention-getting.

Fortunately, he pointed out to Cocconi, there is such a rendezvous in the radio spectrum, and this became the kernel of the joint proposal that the two men discussed in Moscow and then sent off to *Nature* from Cocconi's new home in Switzerland. 'Just in the most favoured radio region,' they wrote, 'there lies a unique, objective standard of frequency, which must be known to every observer in the universe: the outstanding radio emission line at 1420 Mc/sec [21 centimetres].'

Copies of their proposal, distributed in early August of 1959, went to many of those who, they thought, would be receptive. In particular they sent it to P. M. S. Blackett, the noted British physicist, in the hope that, as a friend of Morrison, he could see that it was published in *Nature* despite its sensational content. *Nature* seemed the logical choice, since it had so often been the vehicle for bold new ideas (as in the debate on fossil signs of life in meteorites). The article was published in the issue of 19 September and precipitously brought the debate into the open.

After reviewing some of the initial arguments in Cocconi's first letter, the two physicists pointed out that, while the lifetimes typical of citizens elsewhere in the universe are unknown, it is not unreasonable to suppose that some endure 'for times very long compared to the time of human history, perhaps for times comparable with geological time'. Such beings must have achieved levels beyond the reach of our most imaginative speculators and should be awaiting the appearance of intelligent life in our solar system. 'We shall assume,' they wrote, 'that long ago they established a channel of communication that would one day become known to us, and that they look forward patiently to the answering signals from the sun which would make known to them that a new society has entered the community of intelligence.'

What would be the nature of signals from such an inquiring civilization? Cocconi and Morrison suggested that they might come as pulses, transmitted at a rate of about one per second, with prime numbers or 'simple arithmetical sums' as a device for attracting attention. Because we have to look out 15 light years to find seven stars with luminosities and expected lifetimes comparable to that of our sun, we must think in terms of travel times of many years for messages between civilizations. One light year is the distance travelled by light waves or radio waves in a year, at roughly 186,000 miles a second, and therefore it would take 10 years for a radio message to reach a star 10 light years away and 20 years for an exchange of messages (see the drawing on p. 203).

In this situation a message from a distant civilization would logically be extremely long, taking perhaps years for its transmission. As elaborated by Morrison in subsequent discussions, such a lengthy message would be interspersed with attention-getting signals and 'language lessons'. The bulk of the message, however, would be a detailed description of the transmitting civilization – enough to keep scholars at the receiving end busy during the years required for a reply and further message exchanges. As a possible timetable he suggested one-minute periods for the calling signal, followed by ten minutes for a language lesson and then fifty minutes of information comparable to that found in an encyclopedia, whereupon there would be another minute of calling, ten minutes of language instruction and fifty minutes of the encyclopedia. The 'language' would, of course, not be spoken but would be a system of symbols designed to convey information and ideas. The cycle of language lessons might be completed every ten hours, whereupon it would repeat, but he said it could take fifty years to run through the encyclopedia. Interspersed within the transmissions might be occasional instructions on how to build better receivers, transmitters and so forth. The transmitting civilization would presumably tape-record its messages and bring them up to date from time to time.

The signals would have to be directed at many stars for enormous periods of time before being likely to bear fruit. Cocconi and Morrison said they did not believe a highly advanced society would find it too burdensome to beam signals at, say, 100 nearby suns that seemed promising abodes of evolving life. An antenna like the huge one projected by the Navy for Sugar Grove, West Virginia, they said, would be able to send messages to planets 10 light years away, assuming a receiving antenna of the same size at the other end, and using transmitters no more powerful than those already available on earth.

The dish then under construction at Sugar Grove, some 30 miles northeast of Green Bank, was to be fully steerable, like that at Jodrell Bank, but it was to be 600 feet in diameter compared to the 250-foot width of the British antenna. One of its tasks, as later reported in the press, was to eavesdrop on Soviet domestic radio traffic reflected back to earth by the moon. The project later ran into seemingly insuperable engineering difficulties. To be effective, such an antenna must retain its parabolic shape to within about one-sixteenth of the wavelength to be observed. This permits almost no sag at all, as the dish is swung and tilted to scan various parts of the sky, and demands a frame of great strength. It was finally found that a movable array weighing some 36,000 tons would be necessary to meet these requirements, yet the supporting structure already being built was not adequate for such a weight. Hence, the project was finally abandoned, much to the dismay of those who looked forward to its use as an extremely powerful receiver for scientific research – including, perhaps, the search for signals from other worlds.

Morrison and Cocconi pointed out in their article that, if another civilization had already made contact with sister worlds elsewhere in the galaxy, it would be far more persistent and patient in its search for emerging technologies.

The two men sought to narrow down the search further by selecting suitable targets for a listening programme. They calculated that there are some 100 stars of appropriate type within 50 light years. Of the seven that are within 15 light years, three (Alpha Centauri, 70 Ophiuchi and 61 Cygni) are viewed from the earth against the back-drop of the Milky Way so that the 21-centimetre emissions coming from beyond them are forty times more intense than in other portions of the sky. Hence signals from near those stars, on that wavelength, could be observed only if they were extremely strong. The other four (Tau Ceti, O_2 Eridani, Epsilon Eridani and Epsilon Indi) are suspended against less noisy back-drops in the southern half of the sky. At least two in the first group were, in the view of some astronomers, poor candidates on other grounds. Alpha Centauri, third brightest of all stars and our nearest neighbour, seems to be in a three-star system, one of whose components is very much like our sun; but, as pointed out by Su-Shu Huang, the system is probably too young for life to have evolved. The other, 61 Cygni, is also in a multiple star system, as noted in Chapter 6, and planetary orbits in such systems are not apt to be stable enough to foster a uniform climate and, hence, the emergence of life (see Chapter 7).

One phenomenon that might aid in distinguishing signals generated on a planet from the general din of emissions on 21 centimetres, the two physicists said, was the effect of orbital motion of the transmitter on the frequency of the signal reaching the solar system. This referred to the ubiquitous Doppler effect. They calculated that such relative motion could alter the typical hydrogen frequency as much as 300 kilocycles, up or down. The earth, for example, travels around the sun at some 67,000 miles an hour. If a distant civilization monitored our radio signals from directly above or below the solar system, there would be no Doppler effect, any more than there is relative motion between one's eyes and the edge of a spinning record-player when one looks down at it from above. However, if one views the record-player sideways, the relative motion is maximal. The chances are small that the orbit of a distant planet would be so oriented as to eliminate all Doppler effect.

In conclusion Morrison and Cocconi urged the readers of *Nature*, who include most of the world's scientific élite, not to dismiss this analysis as a form of science fiction.

> We submit, rather that the foregoing line of argument demonstrates that the presence of interstellar signals is entirely consistent with all we now know, and that if signals are present the means of detecting them is now at hand. Few will deny the profound importance, practical and philosophical, which the detection of interstellar communications would have. We therefore feel that a discriminating search for signals deserves a considerable effort. The probability of success is difficult to estimate; but if we never search, the chance of success is zero.

14
Project Ozma

In one of those remarkable coincidence that occur repeatedly in scientific history, at the time when Cocconi and Morrison were suggesting a search for intelligent radio signals on 21 centimetres, a plan to do so was quietly in preparation. Such is the logical flow of scientific thought that it leads to simultaneity in many such developments.

The search was being prepared by Frank D. Drake, a young astronomer at the newly built National Radio Astronomy Observatory at Green Bank, West Virginia. From boyhood Drake had been captivated by the idea that there might be life in other worlds and when, as an undergraduate at Cornell in 1951, he heard Otto Struve lecture on astronomy and the significance of the slow-rotating stars as evidence for a multitude of solar systems, Drake found that such ideas were shared by a man of the highest distinction.

After being graduated in 1952 from Cornell, home base of Morrison and Cocconi, he fulfilled his three years' military service as Electronics Officer aboard the heavy cruiser USS *Albany*, acquiring practical experience that would stand him in good stead when it came to attempting longer-range communications than any contemplated by the Navy. After his discharge he took his doctorate in astronomy at Harvard. Concentrating more and more on the new and exciting field of radio astronomy, he took a job, on the side, working for Harold I. Ewen, co-discoverer of the 21-centimetre emissions, who had gone into business on his own as a founder of the Ewen-Knight Corporation, one of the burgeoning electronics firms in the Boston area. Drake became intensely interested in the 21-centimetre signals as a wedge for prying into various astronomical problems and he is credited with being the first to propose, on the basis of radio observations, that Jupiter has belts of trapped, gyrating particles like those girdling the earth, only 1,000 times more intense –

sufficiently so, in fact, to produce strong synchrotron radiation.

These various lines of activity, at Harvard and then at Green Bank, repeatedly directed his mind toward the question of communication with distant worlds. When he arrived at Green Bank in April 1958, as one of the initial staff members of the observatory, he found it a scene of mud and snow with no permanent buildings. The observatory offices were in a farmhouse. Yet there was something remarkable about this national listening post on the universe, even in its formative stages. The site was in a valley partly sheltered from extraneous, man-made radio signals by flanking mountains. West Virginia had passed a Radio Astronomy Zoning Act to bar producers of undesirable radio interference from moving into the area. It was said to be the first law of its kind ever passed, protecting, as well, the nearby Navy site at Sugar Grove. It applied only to unlicensed radio activity. To curb licensed forms of interference a national 'radio quiet zone' 100 by 120 miles in area was established under the Federal Communications Commission.

Another fear was aircraft interference. The danger was spelled out in the journal *Science* by Richard M. Emberson, director of construction at the site. The metal surface of an aeroplane, he said, can mirror man-made radio signals down into a radio telescope, even though the transmitter is hidden behind mountains. If the plane itself is transmitting, 'the situation becomes quite serious'.

The transmission may not be directly on the frequency of the cosmic signals being observed, yet, he said, their side effects can mask the faint emissions from space. In particular, he noted, the 21-centimetre emissions from hydrogen 'lie in a band assigned to aviation purposes. Thus, for example,' he continued prophetically, 'the spurious signals from flying aircraft in the neighbourhood of Green Bank will interfere with this important astronomical observing band, even if the aircraft transmitter's primary frequency is carefully set away from the hydrogen frequency.'

Eagerness to observe on the magic wavelength of 21 centimetres had been a major factor in persuading the government of the need for a national observatory and, as noted in the last chapter, the first big antenna at Green Bank was designed with such observations in mind. At about the time that this dish was completed, in March 1959, Drake went to a nearby grill facetiously called 'Antoine's' for a quick lunch with Lloyd V. Berkner, acting director of the observatory and head of Associated Universities. The latter was a teaming-up of nine North-eastern universities whose first job had been to establish the Brookhaven National Laboratory on Long Island, home of the world's most powerful atom-smasher. It now had the added task of setting up the National Radio Astronomy Observatory, the government's role in the two projects being justified on the ground that, while scientifically important, they were too costly to be undertaken by any one private institution. Berkner, who had gone to the Antarctic with Admiral Byrd and been a pioneer in radio science, had risen to world prominence as a science administrator. It was he, for example, who first proposed the International Geophysical Year of 1957–58. He was not one to shrink from unusual schemes, and Drake knew it.

Drake's reasoning, in his presentation to Berkner, was not unlike that set forth soon afterward by Cocconi and Morrison in their letter to *Nature*. For thousands of years man had speculated about the possibility of life in other worlds without any hope of verifying the hypothesis. The question had seemed utterly academic. But now, Drake pointed out, the situation was changing rapidly. The reception by Millstone Hill of radar echoes from Venus (later shown to have been announced prematurely) demonstrated the enormous enlargement of man's capabilities. Such amplification devices as the maser and parametric amplifier had been increasing the sensitivity of radio telescopes at a rate of 50 per cent a year. A maser had already been installed at the focus of the 50-foot dish atop the Naval Research Laboratory near Washington, resulting in a sensational increase in sensitivity. A similar device could clearly be used on the 85-foot dish towering above the valley where Drake and Berkner were munching hamburgers. The new dish stood on a three-legged structure that held it aloft like a gigantic robot warrior presenting its shield to heaven.

Drake proposed listening on 21 centimetres more on practical grounds than as an outcome of any attempt, like that of Cocconi and Morrison, to find a 'uniquely rational' channel that would be chosen by a highly advanced society. Observations on that wavelength were, as cited earlier, a prime goal of the 85-foot antenna, and one of the possibilities that excited Drake was that the so-called Zeeman effect could be found in the 21-centimetre line of the radio spectrum. In 1896 a Dutch physicist named Pieter Zeeman had discovered that, when light is emitted by atoms in a magnetic field, the lines of its spectrum are split, each part being polarized. The extent of the effect indicates the strength of the magnetism producing it; hence the Zeeman effect has been used to chart the magnetic fields of the sun and other celestial objects. Could the 21-centimetre line of the radio spectrum be exploited in the same way?

Because extremely sensitive receivers were needed for this conventional investigation, the added cost for equipment to listen for intelligent signals would be only about $2,000, although there would also be operating and manpower expenses, as well as diversion of the much-sought-after antenna from other work. Berkner knew an attempt to intercept signals from another solar system would be scoffed at by some conservative scientists. Nevertheless he encouraged Drake to go ahead, and soon thereafter, at a meeting of the observatory's trustees, he described the project to Otto Struve. A few weeks later, on 3 May 1959, Struve was named the observatory's first full-fledged director and the programme, to a large degree, became his responsibility. It was a natural culmination to his lifelong interest in the possibility of extraterrestrial life. While growing up in Czarist Russia he had read, with his father, the speculations of Percival Lowell regarding life on Mars, and Struve's subsequent work on the slow-rotating stars had persuaded him that planets are plentiful in the universe. In 1960 he estimated that, in the Milky Way Galaxy alone, there are 50 billion solar systems. As to how many of them have produced intelligent life, he wrote in connexion with Drake's project: 'An intrinsically improbable single event may become highly probable if the number of events is very great.' If the probability of finding intelligent life on a planet at one given time is substantially more than one in ten billion, he said, 'then it is probable that a good many of the billions of planets in the Milky

Way support intelligent forms of life. To me this conclusion is of great philosophical interest. I believe that science has reached the point where it is necessary to take into account the action of intelligent beings in addition to the classical laws of physics.'

The idea that intelligence may play a role in shaping events in the universe, alongside the conventional laws of nature, was indeed a revolutionary one. In fact Struve believed that we are on the threshold of a new view of the universe, as remarkable, in its departure from the past, as was that of the Renaissance and the Copernican revolution.

Drake decided to call his effort Project Ozma, 'for the princess of the imaginary land of Oz – a place very far away, difficult to reach, and populated by exotic beings'. He and his colleagues were hesitant about saying anything publicly about their sensational plan and, although the project was initiated in April of 1959, they kept quiet about it until after publication of the Cocconi–Morrison proposal in *Nature* that fall.

When word did get out, there were hoots as well as cheers from the scientific community. It was somewhat embarrassing to Struve, even though, as former president of the International Astronomical Union and one of the world's most respected astronomers, his reputation was almost invulnerable. Project Ozma, he wrote, 'has aroused more vitriolic criticisms and more laudatory comments than any other recent astronomical venture, and it has divided the astronomers into two camps: those who are all for it and those who regard it as the worst evil of our generation. There are those who pity us for the publicity we have received and those who accuse us of having invented the project for the sake of publicity.'

He conceded that the chances of intercepting signals from beings near the first two or three sunlike stars that are scanned were 'almost zero', but he said the instruments developed for the project would benefit all radio astronomy, and he added in conclusion: 'There is every reason to believe that the Ozma experiment will ultimately yield positive results when the accessible sample of solar-type stars is sufficiently large.'

Thus Struve did not envisage Ozma as a one-time effort, but as a continuing search for life in other worlds, carried out, perhaps, with pauses for technological improvement and reconsideration of methods, but with determination to persist until contact is made.

Drake stated his case for Ozma in *Sky and Telescope*, the magazine published at his alma mater, the Harvard College Observatory. By the most optimistic reckoning, he said, we may expect communication signals from a quarter of all stars. At the other extreme he cited the 'extremely conservative' estimate of Harlow Shapley that intelligent life may exist near but one star in every million. The truth, he suggested, lies somewhere in between. He predicted that within fifty years we will have developed radio technology to the point where further improvements will no longer influence out ability to communicate with other worlds. The limitations, henceforth, will be background noise in space and other natural factors. Since our radio technology is now fifty years old, he said, this means that civilizations characteristically jump, in the brief span of one century, from having no capacity for interstellar communication to having the maximum capability. On an astronomical time scale, he said, 'a civilization passes abruptly from a state of no radio ability to one of perfect radio ability. If we could examine a large number of life-bearing planets,' he continued, 'we might expect to find in virtually every case either complete ignorance of radio techniques, or complete mastery.'

Thus the chances that any of our neighbours are in our state of transition are negligible. As Drake told an assembly of physicists on the eve of his listening effort, 'In view of the continuous formation of stars, there should be a continued emergence of technically proficient civilizations.' Since it has taken life several billion years to evolve to our present level and the life expectancy of our planet and sun, in their present form, is another five billion years or more, the present transitional stage from a primitive state to one of unpredictably sophisticated technology is but a moment on that time scale. Therefore, Drake said, we must expect most societies that have crossed the threshold of civilization to be more advanced than our own.

Shortly after making these remarks, at a symposium on radio astronomy at Colgate University in Hamilton, New York, Drake was ready to begin his experiment. As targets he selected two of the stars most often discussed as candidates: Tau Ceti and Epsilon Eridani. Among the others that he considered were Epsilon Indi and Sigma Draconis, but he found that the former was too far south in the sky, remaining at least 5 degrees below the Green Bank horizon. Sigma Draconis, while very like our sun, was considerably farther away than the others. Tau Ceti is in the constellation Cetus, the Whale. Epsilon Eridani is part of Eridanus, the River. Although both stars are about 11 light years away, Drake calculated that signals from planets near them should be observable with the 85-foot dish if they were generated by a million-watt transmitter operating through a 600-foot antenna like the one then being built by the Navy (and later abandoned) at nearby Sugar Grove. To be detected under these circumstances, the signals would have to be concentrated within a very narrow band of frequencies.

In an effort to eliminate background interference, such as man-made emissions and natural 'noise' sneaking in from other parts of the sky, two receiving 'horns' were rigged at the focus of the great 85-foot dish. One horn was aimed to receive signals from the target star (and its planets, if any) reflected to the focus by the parabolic dish. The other horn was aimed away from the star and did not catch the reflected radio waves, yet, like the target horn, it picked up background noise and stray emissions sneaking in from other parts of the sky. An electronic switch, operating at very high speed, alternately connected the two horns to the receiver. All emissions entering the receiver from the horn aimed away from the star were assumed to be interference and this amount of noise was subtracted automatically from the emissions arriving via the other horn. What remained, it was hoped, would be any signals coming from the distant solar system, plus 'cosmic noise' reflected to the focus from the small region of the galaxy lying directly behind the star.

Another arrangement, designed to eliminate the cosmic noise, was based on Drake's assumption that artificial signals, for maximum efficiency, would be concentrated into a band of frequencies no more than 100 cycles in width. The receiver listened simultaneously on a broad band-width, which was

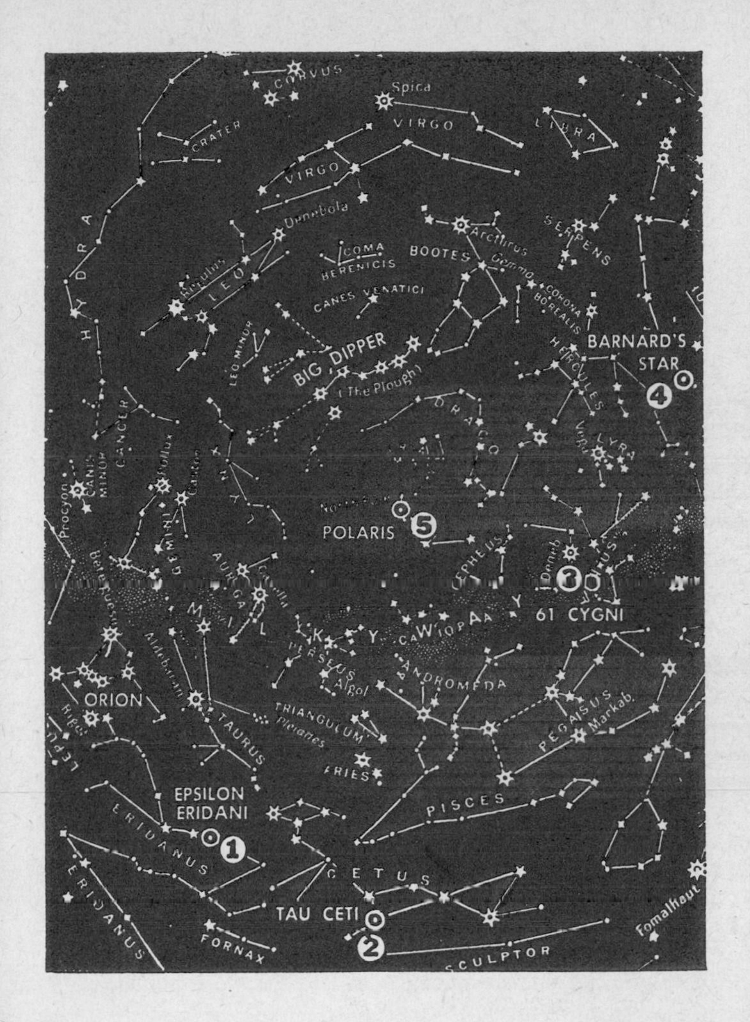

The two stars that figured in Project Ozma were Epsilon Eridani (1) and Tau Ceti (2). The discovery, in 1943, that one star in the two-star system 61 Cygni (3) has an invisible companion led to the search for other such objects. The smallest found so far is in orbit around Barnard's Star (4) which, while one of the closest stars to the sun, cannot be seen with the naked eye. The North Star, Polaris (5), is a variable with a 4-day cycle.

assumed to be almost entirely cosmic noise, and on a narrow band-width. The broad-band noise was then subtracted, electronically, from the narrow-band reception, leaving exposed the signals from life elsewhere.

Of the two new devices for amplifying very weak radio signals, the maser, whose method of operation will be outlined in the next chapter, was the more sensitive but also the more expensive, and had to be kept at temperatures of several hundred degrees below zero, whereas the parametric amplifier was cheaper, simpler, and could be used at room temperature. Drake therefore installed a parametric amplifier to boost the net sensitivity of the horns. This key amplifier was donated by Dana W. Atchley, Jr, president of Microwave Associates Inc., maker of such equipment in Burlington, Massachusetts. Atchley was brimming with enthusiasm about the project and later was one of the few invited to the conference in Green Bank on the problem of establishing contact with other worlds.

Down the line from the parametric amplifier was an oscillator that had to remain constant, in frequency, to within one part in a billion. This was done by using a special quartz oscillator housed inside an oven set for 100°. This oven was itself within another oven, the purpose of the twin ovens being to protect the crystal from even slight variations in temperature.

The equipment was tuned continuously to observe each 100 cycles of band-width for one minute, covering, altogether, about 400,000 cycles of the radio spectrum, centred on the hydrogen frequency of 1420·4 megacycles (1,420,400,000 cycles) per second.

When Drake and his associates took aim at their first star they knew the chances of success were almost nil; yet they could not suppress a certain excitement, if only because this was the start of a search that ultimately could lead to the most revolutionary discovery in the history of mankind.

At roughly four a.m. on 8 April 1960 the dials were set in the control building under the great shield-like antenna, aiming it at Tau Ceti, then newly risen above the mountainous horizon to the southeast. The clockwork was adjusted to keep the antenna pointed at the star as it climbed across the dome of heaven. The receiver was switched on and, for the first time, the two-legged animal that we call man sought to intercept signals from beings of unknown physiognomy in another world.

In the afternoon the star, now visible in daylight, was setting toward the west. No unusual emissions had been observed and Drake decided to shift to the other target, Epsilon Eridani. The output of his equipment was observed in two ways. It was fed into a loudspeaker and also was recorded on moving paper. Shortly after the antenna had been aimed at Epsilon Eridani, and before the loudspeaker had been switched on, the needle inscribing the paper, as Drake put it, 'went bang off scale'. Some very strong signal was coming in! The volume was turned down, and the needle wrote a series of high-speed pulses, at a rate of roughly eight per second, so uniformly spaced that they could only be the product of intelligent beings. When the loudspeaker was turned on, the pulses could be heard, coming in with machine-like precision.

There was, as Drake put it, 'a moderate amount of pandemonium' in the control room. He set all hands to hunting down quirks of circuitry that could account for the pulses, but none came to light. Was it really possible that, on the first day, they had succeeded? The great antenna focused on an area of the sky that, at short wavelengths, was less than a quarter the width of the sun. While there was some leakage into the receiver from other parts of the sky, it was hard to believe that so strong a signal could come from anywhere except the tiny patch of the heavens centred on Epsilon Eridani.

One way to settle the matter was to see if the signal weakened when the telescope was moved so that it did not point at the star. If it did so, the source was probably extraterrestrial, but if the signal retained its strength, the transmitter would be on or near the earth. Before the telescope could be swung away from the star, the signal abruptly stopped of its own accord. It had been received for a total of about five minutes.

Drake was cautious and sceptical. There was no thought of an announcement. Two weeks later the signal again coincided with a listening period and this time its persistent strength, as the antenna was purposely steered off the star, confirmed that the transmissions were of earthly origin. That settled the matter. The pulses were, in fact, heard intermittently for about six months, being observed, as well, by the Naval Research Laboratory. Apparently they were related to a secret military experiment in radar countermeasures, making use of airborne transmitters. They substantiated the warning of the observatory's construction chief, Richard Emberson, that 21-centimetre reception would be particularly vulnerable to aircraft interference.

For a total of some 150 hours, in May, June and July of 1960, the two stars were scanned without further cause for excitement. The search was then suspended, as Drake put it, 'because of the need for the telescope in other projects'. This first phase of Ozma had been intended as a practice run. It was now time to give further thought to such problems as whether or not 21 centimetres really was the uniquely logical channel for the initial establishment of contact between worlds.

To Drake and Struve, however, this was only an intermission. They conceded that the search demanded use of very expensive equipment for a long period. But, Drake said on the eve of Ozma, 'Those who feel that the goal justifies the great amount of effort required will continue to carry on this research, sustained by the possibility that sometime in the future, perhaps a hundred years from now, or perhaps next week, the search will be successful.'

15

Other Channels

While Drake and his colleagues were trying to intercept radio whisperings from two nearby stars, it was proposed that our neighbours in the galaxy may be using a radically different method to establish contact. Ronald N. Bracewell, a leading radio astronomer, argued in *Nature* that radio signals would be an uneconomical way to do this. He noted that the search on 21 centimetres, proposed by Morrison and Cocconi and being carried out by Drake, was aimed at the nearest stars not disqualified by factors inimical to life. But, he said, 'do we really expect a superior community to be on the nearest of those stars which we cannot at the moment positively rule out? Unless superior communities are extremely abundant, is it not more likely that the nearest is situated at least ten times farther off, say, beyond 100 light years?'

If that is so, it means, that after ruling out unlikely candidates, we must still scan one thousand stars to find the highly advanced civilizations that we are looking for and that civilization, in turn, must direct its calling signal at one thousand stars in the hope, ultimately, of finding someone else. 'Remember,' he said, 'that throughout most of the thousands of millions of years of the Earth's existence such attention would have been fruitless.'

It was more logical to assume, he said, that superior civilizations would send automatic messengers to orbit each candidate star and await the possible awakening of a civilization on one of that star's planets. Such a messenger would resemble some of our space probes in being powered by light from the sun to which it was assigned. However, it would be far more sophisticated in design, equipped, as Bracewell explained later, with a miniaturized computer perhaps no larger than a man's head, yet with some characteristics of an intelligent being.

If we contemplate the resources of biological engineering, which we have not begun to tap yet, [he said in a 1962 lecture at the University of Sydney] it is conceivable that some remote community could breed a subrace of space messengers, brains without bodies or limbs, storing the traditions of their society, mostly to be expended fruitlessly but some destined to be the instruments of the spread of intragalactic culture.

Such a messenger, said Bracewell in his *Nature* article, 'may be here now, in our solar system, trying to make its presence known to us.' This dramatic suggestion could not be dismissed as crackpottery, for Bracewell was a respected scientist, co-author of a standard text on radio astronomy and now an associate professor on the faculty of Stanford University in California. Because a messenger might have to wait millions of years before performing its task, it would have to be heavily armoured against radiation damage and meteor impacts, Bracewell said. It would be placed in an orbit lying within the habitable zone surrounding the target star (as defined, for example, by Huang in Chapter 7). The messenger would keep watch for the narrow-band radio emissions announcing that a civilization on a nearby planet has reached the communicative stage. The messenger might then repeat such signals back to the planet. When the civilization awoke to the significance of these strange echoes and itself repeated the signal, the messenger would know it was recognized and begin communicating. 'Should we be surprised,' said Bracewell, 'if the beginning of its message were a television image of a constellation?'

An advantage of this procedure, he pointed out, was that it would lead to use, by the messenger, of a frequency that could penetrate the atmosphere of the new civilization. He pointed out that strange radio echoes have been observed in the past. In 1927, 1928 and 1934, Carl Störmer of Norway and others reported hearing radio signals from Holland repeated from three to fifteen seconds after their original transmission. Some thought they were being reflected from clouds of ionized gas scattered around the solar system, but, as Bracewell pointed out, they were never satisfactorily explained.

Actually, one would not expect such variation in echo intervals if the return signals were being sent by a single messenger. In commenting on Bracewell's idea others have argued, as well, that there are better ways of attracting attention than simply by echoing signals. Nevertheless, Bracewell urged that peculiar radio transmissions of any sort should not be ignored as were, for a time, the remarkably powerful emissions from Jupiter. The latter, he noted, represent an output of one billion watts per megacycle, yet they were not recognized until 1955, when Bernard Burke and Kenneth Franklin of the Carnegie Institution first reported the phenomenon. The source, although clearly not an intelligent species, is still a subject of scientific debate.

It was also possible, Bracewell said, that the probes, 'sprayed' towards nearby stars by a superior civilization, might merely report back when they heard signs of intelligent life, using a star-to-star relay system for more efficient communications. If such a superworld is 100 light years away, the announcement that we have reached the communicative stage has completed only part of its homeward journey. Since radio waves travel at the speed of light, the 100-light-year journey will take them 100 years. The message may include samplings of what was heard, giving the recipients a strange picture of our civilization. Thus, as George A. W. Boehm put it in *Fortune* magazine, we may be assessed initially in terms of Amos 'n Andy, the crooning of Kate Smith, or Gabriel Heatter booming out: 'There's good news tonight.' Soon afterward the gloomy broadcasts of Raymond Gram Swing, on the eve of World War II, will reach our distant neighbours and, as Boehm suggested, they might even decide that the end of civilization on earth is at hand and 'turn the knob to another planet'.

Bracewell said we should not expect that many worlds are seeking to make contact with us, since supercivilizations are probably already linked in a 'galaxy-wide chain of communication'. This federation of intelligent beings in many solar systems would allocate sectors of the galaxy for search by individual civilizations: 'Our impending contact cannot be expected to be the first of its kind; rather it will be our induction into the chain of superior communities, who have had long experience in effecting contacts with emerging communities like ours.'

On 15 April 1961, less than a year after the appearance of Bracewell's proposal, *Nature* published one by two authors arguing for light signals. It will be recalled that, according to

Drake's calculations, to cover the immense distances between solar systems the energy of radio signals would have to be concentrated within a band of frequencies not much wider than 100 cycles per second. One of the great advantages of radio waves for communications is that, by means of various kinds of oscillator, they can be generated in very narrow band-widths. However, until the invention of the maser and its optical counterpart, the laser, there was no known way to do this in that portion of the spectrum occupied by light. Even the most brilliant lamp emitted an inconsequential amount of energy within any narrow part of the spectrum, and therefore the use of light for long-distance signalling was out of the question.

All our conventional light sources, be they electric bulbs, arc lamps or candle flames, produce light by the heating of some material. What happens on the atomic level is that heat stirs up the electrons which, at random times, shed their excess energy in the form of light waves. This haphazard process in billions of atoms produces a wide range of wavelengths. Furthermore, the waves are not 'coherent'; that is, they are chaotically jumbled; there is no symmetrical train of waves that could be modulated to carry information. It is because the waves generated by a commercial radio station are coherent that its output can be modulated to carry a Beethoven symphony or a human voice.

Discovery of the maser and laser grew out of speculation by a number of scientists, including Charles H. Townes at Columbia University in New York, N. G. Basov and A. M. Prokhorov in the Soviet Union, and Joseph Weber at the University of Maryland, concerning possible ways to 'organize' the excitation of atoms and their subsequent radiation of energy. An atom or molecule can become excited either in response to a wave at its resonant frequency or by the action of heat-induced collisions. Once it is excited, there are two ways in which it can radiate a wave at its characteristic frequency. One is by spontaneous emission, at an unpredictable time of the atom's own choosing, so to speak. This is the manner characteristic of electric lights. The other is by 'stimulated emission', which occurs when an excited atom or molecule encounters a wave at its resonant frequency. Since the atom is already excited, the new wave jars loose from the atom another wave of that frequency and the two waves go sailing off, hand in hand. More scientifically, they are 'in phase', like soldiers marching in step. They are therefore the stuff of which a smooth wave-train is made – a coherent wave, suitable for message-carrying.

The first maser was developed in 1954 when Charles H. Townes, James P. Gordon and Herbert J. Zeiger at Columbia University found a way to assemble excited ammonia molecules in a resonant, metal-walled chamber. This could be done because excited ammonia molecules are repelled by an electrostatic field, whereas such molecules in the lower energy state are attracted by it. The experimenters sent a stream of low-pressure ammonia gas through an electrostatic field, segregating the excited molecules, which were then directed into the metal chamber. There they waited, aquiver with excitement for a wave of the right frequency. (For ammonia, it is 24,000 megacycles, which lies in the microwave, or radar, region of the spectrum.) When such a wave entered the chamber and encountered one of the molecules, it released an identical wave, thus producing two waves (or photons) from one. These, in turn, interacted with other molecules, releasing more waves which bounced between the walls of the chamber, constantly producing more and more waves on that same frequency. It was 'microwave amplification by stimulated emission of radiation', and the device came to be known, from the initial letters of that formidable title, as the 'maser'.

While such early masers were able to amplify microwave signals to an extent never previously possible, they were useful only in a very narrow frequency range centred on 24,000 megacycles. The next discovery, thanks largely to the work of Nicolaas Bloembergen at Harvard, was that similar amplification on a number of frequencies could be achieved by increasing the energy states of free electrons in a crystal structure. For example, some masers were made with a silicon crystal containing occasional atoms of phosphorus. The latter fit into the crystal lattice so that one electron from each atom is free and available for excitation by microwaves on frequencies other than that to which these electrons are resonant. Once excited, the electrons will amplify waves on their own frequency.

In 1958 Townes and Arthur L. Schawlow showed how it would be possible to apply the maser principle to light waves and thus produce a unique kind of light. In the 15 December issue of the *Physical Review*, they proposed that a resonant chamber be built with mirrors facing each other at each end, the distance between them being thousands of times greater than the wavelength to be generated, but nevertheless a multiple of half that wavelength, thus permitting light to resonate between the mirrors. Light waves generated along the axis of the maser would travel back and forth between the mirrors, constantly growing in intensity as more electrons in the central crystal gave off their stored energy. A partial transparency of one mirror would allow some of these waves to escape in a beam that was not only very directional and intense, but was concentrated in a narrow frequency band and was coherent.

The first maser to operate within the frequency range of visible light was dependent on a pencil-like rod of synthetic ruby whose ends were machined with great precision and silvered to provide mirrors facing each other along the axis of the rod. Around the rod was a coil of xenon flash-tube that generated an intense light, serving to excite electrons within the crystal. Ruby consists of aluminium oxide with a few of the aluminium atoms replaced by chromium atoms. The more of the latter there are, the deeper the red of the ruby. Within the lattice of interlocking atoms that form a ruby crystal, these chromium atoms have extra electrons that are free of the lattice and hence can be excited to higher energy states by the action of ordinary light. Light at the characteristic wavelength of a particular ruby crystal, usually near 7,000 angstroms, produces a cascading of the excited electrons and a shower of additional waves of precisely the same colour. (An angstrom is $\frac{1}{100,000,000}$th of a centimetre.) In the first lasers, only those waves directed along the axis of the rod could survive and gain in intensity. Those aimed obliquely flew out of the crystal and were lost.

The successful production of such a device was announced in July 1960 by Theodore H. Maiman of the Hughes Aircraft Company, and in rapid succession it was found that other crystals would produce light at various wavelengths. It was characteristic of these early devices that they had to operate in pulses, but in February 1961 Bell Telephone Laboratories announced the development of a continuously operating type that used a mixture of helium and neon gas instead of a ruby. Such devices, whether dependent on a crystal or a gas, soon came to be known as lasers (for 'light amplification by stimulated emission of radiation').

There followed a scramble to use them for a wide variety of purposes, including pinpoint destruction of tissue in delicate surgery and for various forms of communications. The available frequencies spread across the infra-red and visual parts of the spectrum. On 9 May 1962 a laser beam generated by a ruby crystal the size of a 6-inch pencil was used by a group at the Massachusetts Institute of Technology to illuminate a small patch of the darkened moon. The stark lunar landscape glowed sufficiently for the reflected light to be detectable by instruments in Massachusetts.

By this time the increase in laser capabilities had already stimulated the thought that such a device might be chosen by those in other worlds seeking to attract our attention. It is not surprising that this was first proposed by Charles Townes, 'father' of the maser and laser. It was he and Robert N. Schwartz who co-authored the 1961 article in *Nature*. Both were by then at the Institute for Defense Analyses in Washington, where Townes was Director of Research. He was soon to become Provost of M.I.T.

It seems to have been a 'historical accident', they said, that lasers were not discovered before radio as a means of long-range communications. Another civilization 'might have inverted our own history and become very sophisticated in the use of optical or infra-red masers rather than in the techniques of short radio-waves'. In our case, however, lasers are at a rudimentary stage of development. 'No such operating device was known a year ago,' they wrote, yet the technology in this field promises to advance at great speed, with major gains in the next decade. Already, they added, we are not far from the development of lasers able to communicate on wavelengths of visible light, or in adjacent portions of the spectrum, 'between planets of two stars separated by a number of light years'. Furthermore, they said, it may be possible that our present-day telescopes and spectrographs might be able to pick up signals from another civilization not much more capable than we are.

They discussed two methods of interstellar communication by laser which they designated System A and System B. The former would pass the beam of a single laser through an optical system with a 200-inch reflector, comparable to that of the world's largest telescope on Mount Palomar in California. This would produce a beam that, to start with, was 200 inches wide, instead of the very narrow beam typical of a laser; but the spread of the beam would be sufficiently reduced so that, millions of miles away, it would be much narrower than if such an optical system were not used.

The other scheme, System B, would use a battery of 25 lasers without any optics. They would be clustered to produce a beam initially only 4 inches wide; but the spread of the beam, slight as it was, would be five times that of System A. However, unless the laser and the huge telescope of System A were lifted above the atmosphere, the advantage of its well-disciplined beam would be lost to the light-scattering effects of the air. Both systems would use 10-kilowatt lasers generating a wavelength of 5,000 angstroms.

Schwartz and Townes pointed out that a few years ago the idea of operating a laser from a space platform or from a high-altitude balloon would have seemed out of the question. Now orbiting observatories are in preparation, and balloon-borne telescopes are coming into their own. In 1963 Stratoscope II, a 3-ton instrument with a 36-inch mirror, demonstrated that such remote-controlled telescopes can be operated with great precision when suspended from balloons. However, it was a big jump from a 36-inch telescope to one of 200 inches – of which there is only one in existence. Even though System A could deliver to the targets light 100 times as intense as System B, assuming the equipment were carried aloft, the two scientists conceded that at present System B was the only practical one for earth's inhabitants.

Under favourable seeing conditions, they calculated, a person could view the signals from System A at a distance of 0·1 light years with the unaided eye. With binoculars the range would be some 0·4 light years. For System B the distances would, in each case, be one-tenth as large. If, however, a telescope like that on Mount Palomar were in use at the receiving end, System A could be observed visually at 10 light years and System B, at the same range, would show up on photographic plates exposed for one and a half hours. Thus they demonstrated that even our primitive laser technology is nearing the point where it can send observable beams to the nearer candidate stars.

What, then, should we look for? At a distance of 10 light years, the earth and the sun would appear only one-half second of arc apart. Most telescopes could not separate objects so close together, even if they were above the turbulent atmosphere, although the 36-inch mirror in Stratoscope II is designed to resolve objects only one-tenth of a second apart. Hence, the transmitting civilization must see to it that its laser signals are not hidden by the brilliance of its nearby sun. One way to avoid this, Schwartz and Townes pointed out, would be to select a wavelength of light that is absorbed by stellar atmospheres.

When one looks at the sun, even through a pocket spectroscope, its brilliant spectrum of colours is subdivided by a series of black lines. It is as though some physics teacher had drawn the lines there, with precise neatness, to aid his students. But the lines are 'drawn' by nature. As light from the sun passes through the solar atmosphere on its way to the earth, gases in that atmosphere absorb the light at their characteristic wavelengths, leaving black lines in those parts of the spectrum. Two of the most prominent lines, in this respect, are those absorbed by calcium in the violet (at wavelengths of 3933 and 3968 angstroms). Each line is several angstroms wide and, at those wavelengths, the light of the sun is dimmed some ten-fold. If a laser transmitted light in an extremely narrow line at the centre of such a broad 'black' line in the star's spectrum, astronomers on a distant planet, scanning the light of various stars, might recognize it as artificial. They could then examine this light more carefully to see if it was coded, in some way, to carry a message.

The narrower the band of the emitted light, the brighter and more obvious it would be, so long as astronomers at the other end were capable of detecting such narrow bands of emission. Our present capabilities in optics limit us, in general, to the detection of spectral lines representing a spread of at least 0·01 angstroms in wavelength. One problem pointed out by Schwartz and Townes was that the wavelength spread of a laser beam, though narrow to begin with, could be broadened, and the signal therefore dimmed, by shifts in observed wavelength caused by changing relative motion between transmitter and receiver. The most likely cause of such Doppler effects would be the spin of the home planets of both the receiving and sending civilization. The effect could become significant when film was exposed for several hours to pick up faint lines in the spectrum. It was therefore proposed that the transmitting civilization might adjust its wavelength to allow for the spin and orbital motion of its planet and that those on the receiving end might make similar allowances for their own motion.

The two scientists in effect apologized for questioning the uniqueness of 21-centimetre radio waves as the logical channel. 'It may be both an encouraging enlargement of possibilities, and at the same time an unwelcomed complication,' they said. 'The rapid progress of science implies that another civilization, more advanced than ourselves by only a few thousand years, might possess capabilities we now rule out,' they said. 'They may have already been able to send us an exploratory instrumented probe.' (See the discussion of interstellar travel in Chapter 16.) 'Since none has yet been seen,' they continued, 'perhaps it would be appropriate to examine high-resolution stellar spectra for lines which are unusually narrow, at peculiar frequencies, or varying in intensity.'

Those who leaned toward radio waves as the best vehicle for interstellar signalling found drawbacks to the laser proposal. Among the radio protagonists was Bernard M. Oliver, Vice President for Research and Development of the Hewlett-Packard Company, makers of electronic measuring devices in Palo Alto, California. In the *Proceedings of the Institute of Radio Engineers*, he discussed the great variety of new capabilities that lasers have provided; but he concluded that, before such devices can be useful for signalling to other solar systems, 'a tremendous increase' in laser output must occur.

Drake came to a similar conclusion. He argued, as well, that mirrors used to collect light signals must be far more precisely made than those in radio telescopes, since the permitted error is controlled by wavelength. The production of a mirror 200 inches in diameter for the Mount Palomar telescope was a feat of optical technology that has never been repeated, yet the dish of the Jodrell Bank radio telescope is 3,000 inches wide, and one constructed in a bowl-shaped valley in Puerto Rico is 24,000 inches across. Drake sought to analyse the problem in terms of physical laws that are universal and would therefore influence engineers on any planet of the universe. Thus he argued that the choice of wavelength is partly controlled by the flexibility of metals and other solids that might be used in building mirrors for radio or optical waves. Such flexibility, determined by the inherent binding properties of solids, points to the longer, or radio, wavelengths as those for which the largest dishes could be built. The limitations we encounter in antenna structures, he said, 'are not peculiar to ourselves, but are common to all civilizations.'

Also universal are two of the mysterious 'constants' that affect all forms of electromagnetic radiation, be it in the light or in the radio parts of the spectrum. They determine the extent of the 'quantum' noise that results from the very nature (or quantum behaviour) of light waves. The extent of this noise, at any frequency, is equal to the frequency multiplied by the Planck constant and divided by the Boltzmann constant. This law must be known to every physicist in the universe, be he a hundred-legged blue sphere or a two-legged cylinder, like Frank Drake and the rest of us. The effect of the law is to point toward lower frequencies and longer wavelengths as the channels most free of quantum noise. Thus, as pointed out by Oliver, while a radio system has far more beam spread than one using light, the quantum noise, per cycle, at optical frequencies is thousands of times greater than at the proposed radio frequencies.

The optimum frequency for communication with any one star is also determined by the competing background noise from that part of the galaxy lying behind the star, as seen from the earth. Thus, while the consideration of quantum noise (the inherent noise in all electromagnetic radiation) favours low frequencies,

ELECTROMAGNETIC SPECTRUM SUITABLE FOR SIGNALLING

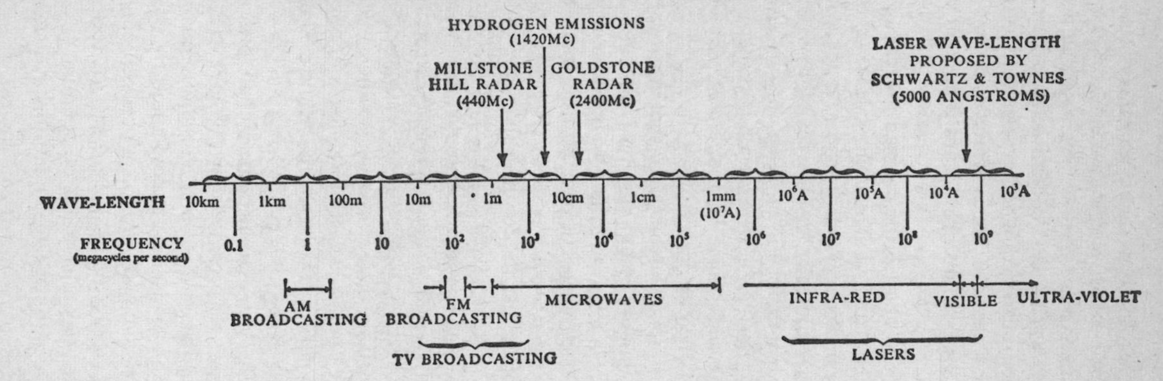

the noise from various parts of the galaxy sets a lower limit on such frequencies. Drake calculated that the most favourable frequency for conversing with any star will lie between 3,700 and 9,300 megacycles (3·2 and 8·1 centimetres). We can already send radio messages out 1,000 light years, he said, bringing some two million stars within our range, whereas, at our present level of technology, laser signals could hardly be seen beyond 10 light years.

Such arguments have not impressed Townes and the other laser adherents. They point out that our present engineering and laser capabilities are a poor guide to those of others more advanced than we. Furthermore, such disadvantages as greater quantum noise, for light waves, may ultimately be outweighed by the narrower and more intense beam of a laser. The truth, according to Townes, is that we are not yet in a position to decide with assurance which is best.

One proposal by Bernard Oliver that at first seemed attractive was to look for signals of extremely short duration and broad band-width. He cited a peculiarity of radio signalling, namely, that the shorter the length of a pulsed signal, the broader the band-width necessary for efficient transmission and reception. This seemed to offer a way to get around the problem of frequency selection. He calculated that, if one could cram 100 million billion watts into a pulse a ten-billionth of a second long, the optimum band-width would span the entire radio spectrum to which the atmosphere is transparent. Assuming that we could detect such very short pulses (although at present we cannot), it would only be necessary to tune the receiver to this very broad band-width and aim the antenna at suitable stars.

Oliver later shelved this proposal because of the effect on such broad-band signals of electrons along their path. While space between the stars is almost a vacuum, there are still enough particles, including electrons, per cubic inch to affect the transmission speeds of radio waves. Through such a medium, waves of higher frequency travel slightly faster than those of lower frequency. The effect on a broad-band signal would be to smear out the duration of each pulse until it became undetectable.

Oliver then proposed that a huge fan-shaped antenna 20 miles long be built across the Equator so that the earth's spin would daily sweep its aim across almost every star in the sky. He conceded that the observing time on each star would only be about one second per day and that a large number of stars would be scanned simultaneously. His point was that this gross scanning might disclose a target worth closer scrutiny.

There were other suggestions. M. E. J. Golay of the Technische Hogeschool in Eindhoven, the Netherlands, cited the disadvantage of listening directly on 21 centimetres, since this involved competing with the natural hydrogen emissions. He suggested that better locations for a search were at half or double the hydrogen frequency. Drake then picked up this idea of examining multiples, or 'harmonics', of the hydrogen frequency, which had also been suggested by Sebastian von Hoerner. Drake said our distant friends may have chosen the harmonic closest to the minimum-noise frequency applicable to their star. Here, in effect, was a refinement of the Cocconi–Morrison proposal. But was it too refined to be 'uniquely logical'?

Philip Morrison noted that the really logical method may be by 'Q' waves 'that we're going to discover ten years from now!' Or, he said, a way may be found to handle neutrinos as message carriers. While these elusive and paradoxical atomic particles travel at the speed of light and can go right through the heart of a planet, no one knows an easy way to generate them, detect them, or modulate them to carry information.

Perhaps the most startling proposal, in the discussion of how other beings could communicate, was that of Leslie C. Edie of Bellmore, Long Island. In the issue of *Science* for 13 April 1962 he suggested that the long-chain molecules now being extracted from certain meteorites (the carbonaceous chondrites) might have been put there by some remote civilization and hurled toward us in great numbers. Might these long molecules contain coded information, he asked?

There are also those who say: 'But what if no one is calling and all are listening?' This does not necessarily mean that all is lost. It might still be possible to detect the more powerful radio emissions produced routinely by an advanced civilization – communication with its interplanetary vehicles, tracking them by radar, and so forth. For example, BMEWS, the Ballistic Missile Early Warning System in Greenland, transmits radar pulses of 5 million watts, powerful enough, probably, to kill a man at short range.

The possibility of interstellar eavesdropping was analysed, at a 1961 symposium of the Institute of Radio Engineers, by J. A. Webb of the Lockheed Georgia Company. The stray signals likely to reach us from a sophisticated society 10 light years away, he found, would probably be undetectable by our present antennas. Matters would be different, however, if the antennas were above the atmosphere, either in orbit or on the moon. Ultimately, he said, antennas 10,000 feet in diameter might be erected in space and these could probably detect a civilization 'tens of light-years' away. Analysis of the receptions would be tedious but, if they were tape-recorded, this could be done by computer. Thus, he said, our technology is probably not yet up to snooping on radio traffic elsewhere in our arm of the galaxy, but eventually may be able to do so.

Frank Drake, following the negative results of Project Ozma, also leaned towards eavesdropping. He reasoned that we are unlikely to call other worlds if we have no real evidence that they exist. Why, then, should we expect them to call us? He proposed that a method analogous to the 'coded pulse' technique used to detect weak radar signals be applied to this problem.

It was the coded-pulse method that enabled Millstone Hill to obtain radar ranges on Venus. A series of pulses is transmitted toward the target and the returning signal is cross-correlated with a record of the original pulses. Although the individual echoes may be lost in the background noise, when they have been aligned with the original pulses they can be superimposed, electronically, producing a composite signal that is evident.

Drake assumed that another civilization like ours would be using a multitude of frequencies for its television broadcasts, communications and other activities. None of these signals would, by itself, be detectable. However, he proposed that a sensitive receiver, aimed at the candidate star, record in segments of 100 cycles at a time, sweeping the 10 billion cycles of the radio spectrum available for such eavesdropping. This would produce records of signal strength in 100 million segments, or bands, of the spectrum. Then a second sweep would be made, producing a second set of records for the 100 million bands. These would be correlated with the first set. In those bands where artificial signals were concealed in the noise, such signals would be atop one another. The noise, being random, would not pile up in this way. The superimposition of all 100 million bands would then produce a different effect than if the distant solar system were not using radio.

On the basis of our heavy use of the radio spectrum, Drake estimated that a radio telescope aimed at another world like our own would encounter artificial signals in about half the bands scanned. Their cross-correlation would make it possible to detect their existence even though the individual signals were 100 times too weak for observation. 'This approach suggests,' said Drake, 'that our civilization may itself be easily detectable.'

A quite different group of proposals has been based on the assumption that a distant civilization would try to attract our attention by means of some 'marker' or beacon. One idea, discussed by Drake and others, was to enclose one's sun in a cloud of material that absorbs some unusual – and hence clearly artificial – wavelength of light. As noted earlier, there are black lines in the spectra of the sun and other sunlike stars, denoting substances in their gas envelopes that absorb certain wavelengths. Suppose that in the light of a distant star we found an absorption line that we could not explain naturally? Drake cited the example of technetium. This element is not found on earth and only weakly in the sun, in part because it is short-lived, in its most stable form decaying radioactively within an average of 20,000 years. As the name of technetium implies, it is observed on earth only where produced artificially. Although seen in the spectra of certain unstable stars, it is not readily detectable in those resembling our sun. Therefore, if we saw evidence of technetium in the light of a sunlike star, it would be cause for further investigation. Drake estimated that such star-marking would require only a few hundred tons of a light-absorbing substance, spread around the star; but the scheme would still place heavy demands on a civilization's technology. Not only would the stuff have to be distributed uniformly, but it would have to be maintained in the face of the steady outflow of gas from the sun – the so-called 'solar wind'. The great attraction of such a scheme is that it would take advantage of the enormous output of radiation from a star, freeing the civilization from the need to generate such vast amounts of energy and letting the star do the work instead.

Philip Morrison discussed converting one's sun into a signalling light by placing a cloud of particles in orbit around it. The cloud would cut off enough light to make the sun appear to be flashing when seen from a distance, so long as the viewer was close to the plane of the cloud's orbit. Particles about a micron (one-millionth of a metre) in size, he thought would be comparatively resistant to disruption. The mass of the cloud would be comparable to that of a comet – some 100,000 billion tons – covering an area of the sky five degrees wide, as seen from the sun. Every few months the cloud would be shifted to constitute a slow form of signalling, the changes perhaps designed to represent algebraic equations.

An inadvertent marker was inherent in the 'far-out' proposal of Freeman J. Dyson, an Englishman at the Institute for Advanced Study in Princeton, New Jersey. In 1960 Dyson discussed how distant supercivilizations may have used their technology. He argued that population growth in such worlds would continue to press the limits of available sustenance, as set forth in the eighteenth-century theory of Thomas Robert Malthus. The limiting factors would be available material and available energy. Both shortages could be met, Dyson argued, by dismembering planets of that solar system and using the material to build a shell completely enclosing the parent star. The entire energy radiated by the star would then become available – 40,000 billion times that which falls on the earth.

He calculated that from the material of Jupiter such a shell could be built with energy equivalent to that radiated by the sun in 800 years – a short period on the time scale under discussion. The shell would be 6 to 10 feet thick, surrounding the star at twice the distance of the earth from the sun. The outside of such a shell, because it was warm, would radiate just as copiously as the central star, he said, but in the infra-red part of the spectrum, centred near a 10-micron wavelength, which easily penetrates the earth's atmosphere. Hence, he proposed a search for sources of radiation on that wavelength.

Readers of *Science* pointed out in letters that such a shell could not survive the gravitational stresses and the tendency of the material to move into the equatorial plane of the shell's rotation. Dyson replied that he envisioned a 'loose collection or swarm of objects travelling on independent orbits around the star'.

When I visited Shklovsky, late in 1962, the Soviet astronomer developed Dyson's idea further. A civilization capable of such feats, he said, would think nothing of allocating one or two per cent of its solar energy to a beacon radiating on a key wavelength, such as 21 centimetres, particularly if the beacon seemed to offer the only chance of drawing the attention of another civilization. Shklovsky was pessimistic as to the density of civilizations in space. They are probably at least 700 light years apart, on the average, he said, since we have not inadvertently stumbled upon the signals from one of them. If a civilization could divert a small percentage of its star's energy into a suitable wavelength, it would be a marker *par excellence*. To find such a beacon in our own galaxy, said Shklovsky, would probably be a long task, since, with the entire sphere of heaven to choose from, it would be hard to know where to start. Instead, he proposed to look at the great spiral nebula in Andromeda, the nearest galaxy like our own. Being two million light years away, the entire galaxy is encompassed within the field of view of a radio telescope. So brilliant a beacon should be evident even at this distance. If one was seen in Andromeda, it would mean there was a supercivilization there two million years ago and this, Shklovsky said, would then justify a patient search of our own galaxy.

His line of thought soon bore dramatic fruit. Early in 1964 one of his associates, Nikolai S. Kardashev, proposed in the scientifically conservative *Astronomical Journal* of the Academy of Sciences of the U.S.S.R. that two heavenly sources of radio emission might, in fact, be beacons of supercivilizations. They had been catalogued as CTA-21 and CTA-102 by the California Institute of Technology and had been singled out the previous year in a survey of 160 such 'discreet sources' (or radio stars) published in the *Monthly Notices* of Britain's Royal Astronomical Society. What had struck the authors was the peculiar shape of the radio-frequency spectra of these sources, compared to those of the others. In both cases their emissions were strongly concentrated towards 900 megacycles. The emissions, Kardashev pointed out, are thus centred in that part of the spectrum best suited to communications across the cosmos. Furthermore, he cited the calculation of his colleague, V. I. Slish, that, unlike the other radio sources, these two were virtually pinpoints in the sky.

To support his argument Kardashev proposed that there are three levels of technology: One is comparable to that of our planet which, as a whole, can generate 4,000 billion watts of power. He classed it a 'Type One' civilization. Then, he said, there are those postulated by Dyson with energy at their disposal equivalent to that radiated by a star – 400 million billion watts. With the earth's annual three to four per cent increase in power production, ours could become such a 'Type Two' civilization in about 3,200 years. Finally he spoke of 'Type Three' civilizations that dispose of energy comparable to that generated by an entire galaxy. His proposal was that peculiar radio sources, such as CTA-21 and CTA-102, might be beacons of Type Two or Type Three technologies.

What created a world-wide sensation was the report by another of Shklovsky's colleagues at the Sternberg Astronomical Institute that he had observed rhythmic fluctuations in the signals from CTA-102. This first appeared quietly on 27 February 1965, in the *Information Bulletin on Variable Stars*, published under the auspices of the International Astronomical Union. The author, Gennady B. Sholomitsky, said that for several months he had repeatedly compared the strength of emissions from CTA-102 with those from a presumably unchanging source (3C-48). He did the same with CTA-21. The emissions from the latter proved to be steady, but those from CTA-102 fluctuated in what appeared a smooth 'sine curve', with peaks at roughly 100-day intervals.

In his report Sholomitsky did not speculate as to the reason for the fluctuations, but on 12 April 1965 the Soviet news agency TASS announced that he had found CTA-102 to be the beacon of a 'supercivilization'. It was evidence, the agency said, 'that we are not alone in the universe.'

This created such a furor that Shklovsky, Kardashev and Sholomitsky held a hurriedly summoned press conference the next morning. Shklovsky criticized the TASS report and said it was 'a little premature' to say the signals were artificial. However, he maintained that this was still at least possible.

Soon thereafter the idea that CTA-102 might be a beacon was dealt a severe blow. Mount Palomar reported that a peculiar, blue-shining object had been found in the precise location of CTA-102. Furthermore, Maarten Schmidt of that observatory had discovered that the spectral lines of its light were so radically shifted towards the red end of the spectrum that the object must be receding from the solar system at some 114,000 miles a second. In our expanding universe this speed can be used as a rough index of distance. It showed CTA-102 to be one of the most remote objects ever observed – apparently one of the newly discovered bodies known as 'quasars', whose inadequately explained brilliance makes them visible, both in light and radio waves, far beyond anything else. It was hard to believe such emissions could be artificial and so the flurry subsided, although the fluctuations that have been observed in the energy output of this and several other quasars remain a mystery.

16

Can They Visit Us?

THE discussion of communications between solar systems has dealt primarily with long-range signalling, apart from Bracewell's proposal for automated messengers. But what about travel between distant civilizations? Should we expect visitors from another world, particularly after they have detected radio emissions revealing the emergence of intelligent life in our solar system? Should we look for evidence of earlier visitations?

While early speculators thought it might be possible to fly to the moon on wings or by balloon, such thinking was terminated by the discovery that the atmosphere thins to almost nothing at a height of about 20 miles. Not until the writings of Konstantin Eduardovich Tsiolkovsky on rocket travel were thoughts of space journeys revived. His classic work, *Exploration of Space by Means of Reactive Apparatus*, was published in Russia in 1896 when Tsiolkovsky was thirty-nine years old. It sets forth many of the problems being encountered now that space travel is a reality. By the time of his death in 1935 at the age of seventy-eight, he had written extensively on the possibility of life in other worlds, sometimes with a formidable imagination. This is reflected in his correspondence with Alexis N. Tsvetikov, then a graduate student at the Biochemical Institute of the Academy of Sciences in Kiev and now in the Department of Biophysics at Stanford University in California. Tsiolkovsky noted in one of his letters to Tsvetikov that in the biochemical sense life on earth 'seems to be a relatively separate unit'. Individuals die, he said:

> However, the total amount of living matter perseveres, and even increases. We can imagine a spherical organism with the cycles of physiological processes closed completely in themselves. Such an organism will be immortal and photosynthetic, and it can develop even a higher consciousness. . . . The main activity of the highest living organisms in the Universe can be also the colonization of other

worlds. Such beings, probably, could not be of spherical form, and they will not be immortal.

In the last months I have often been dangerously ill ...

This letter was written on 2 October 1934, a few months before Tsiolkovsky died. Among his other proposals were life forms composed entirely of hydrogen (reminiscent of the living plasma cloud that, in Fred Hoyle's science-fiction novel *The Black Cloud*, invades the solar system and begins feeding on the sun). Tsiolkovsky even envisioned the ultimate in beings as a disembodied entity, 'living' in space as an almost godlike island of pure consciousness. Because plants lack feeling and provide essential food for higher life forms, Tsiolkovsky saw a place for them in his superior worlds; but he regarded animals as early stages in the painful process of evolution. A superior race, he said, would painlessly eliminate them rather than see them endure the needless sufferings of evolution and the struggle for existence. He argued that, like good gardeners, such superior beings would weed out lower animal species, harmful bacteria and valueless plants, except for laboratory samples.

Tsiolkovsky ended his book *Monism of the Universe*, first published in 1925, with a number of axioms, one of which was that on at least one planet, somewhere, beings have achieved a technology permitting them 'to overcome the force of gravity and to colonize the Universe'. Consequently, he said, 'perfection and dominance of the mind' have been spreading through the cosmos. This was initiated an infinite time in the past and colonization now is the normal manner in which life spreads. Evolution, 'with all its sufferings', is rare, he said. Yet, while he believed in interstellar travel, he considered radio the chief means of communication. In a 1920 letter to the Organization of Young Technicians, he predicted that, 'In the near future short radio waves will penetrate our atmosphere and they will be the main means of stellar communication.'

Tsiolkovsky's far-reaching speculations are somewhat reminiscent of Nikola Tesla. His mysticism was clearly distasteful to the materialistic Soviet state. It is reported that when he died, the government took all of his papers and placed them in the archives of Aeroflot, the Soviet airline. His home, in the city of Kaluga in western Russia, was converted into a museum, but not until the success of the first sputnik in 1957 did he become a national hero.

The surge of excitement that accompanied the dawn of the space age stimulated new interest in the possibility of travel between solar systems. There was talk, for example, of using nuclear-powered rockets driven by the pressure of light itself; but John R. Pierce, Executive Director of the Research-Communications Principles Division at the Bell Telephone Laboratories and one of the nation's leading applied scientists, did a sober analysis of the problem with discouraging results. It was evident that to reach even the nearest sunlike stars within a human lifetime would demand a peak speed close to that of light. Yet it was inconceivable, Pierce said, that a rocket propelled by light could carry enough fuel to reach such a speed, even if there was complete conversion of the fuel into pure energy. If the energy were used, instead, to 'push against' a massive object like the earth, he said, it would be somewhat easier to achieve so high a velocity, but it was hard to see how this could be applied to space travel.

Writing in the June 1959 issue of the *Proceedings of the Institute of Radio Engineers*, Pierce also discussed the possibility of scooping up hydrogen en route as fuel for such a vehicle. Throughout the galaxy there appears to be at least a sprinkling of hydrogen, and in some hydrogen clouds there are thought to be as many as 1,000 atoms per cubic centimetre (thimbleful). Pierce discussed a system using a 'generous' scoop 100 square metres (more than 100 square yards) in area to snatch hydrogen en route. This collecting device, he added, would no doubt consist of force fields, such as magnetism, rather than a material substance. Taking a space ship of 17·5 tons as the smallest conceivable size for interstellar travel, he estimated that the highest speed within reach of such a system would be only 9·3 per cent of the speed of light. 'Clearly,' he said in conclusion, 'it is impossible to attain a velocity close to that of light by using interstellar matter as fuel.'

Another discouraging note was sounded the next year by Edward Purcell, discoverer of the 21-centimetre line in the radio spectrum, in a lecture that he gave at Brookhaven National Laboratory on Long Island. The laboratory, where he was spending a year as Research Collaborator, is the site of the world's most powerful accelerator (or 'atom-smasher'), and his audience consisted largely of physicists. His arguments were elaborated in *Science* a year and a half later by Sebastian von Hoerner, a former associate of Frank Drake at Green Bank who was now at the Astronomisches Rechen-Institut (Astronomical Calculation Institute) in Heidelberg, Germany. Each began by calling to mind the immensity of interstellar distances. If, for example, the sun were scaled down to the size of a cherry-stone, the earth would be a grain of sand three feet away. The nearest star would be another cherry-stone 140 miles away, but no advanced technologies would probably be found nearer than some 25,000 miles.

Purcell pointed out that the strange 'time dilation' predicted by the Special Theory of Relativity would help passengers endure such long journeys if, in fact, they could be accelerated to a speed close to that of light. Let us assume, for example, that a vehicle accelerates at a rate equivalent to the force of gravity at the earth's surface (that is, at a rate of one 'g'). We can be sure that human bodies could withstand such acceleration indefinitely, since the force exerted by gravity on our bodies throughout life is one g.

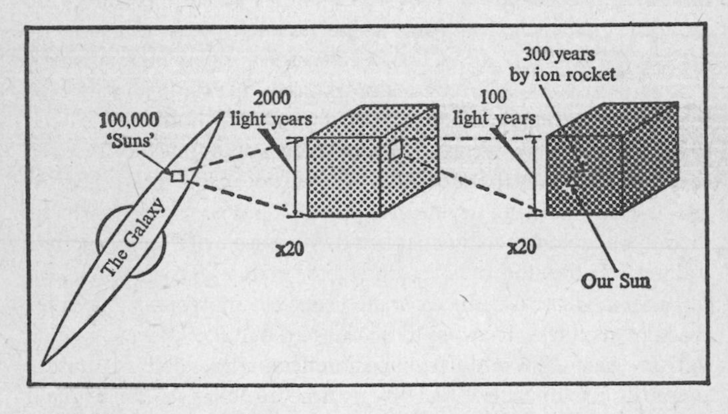

In this diagram, devised by Philip Morrison to illustrate the scale of our galaxy, a small cube of the latter's interior containing 100,000 stars, left, is expanded 20-fold to produce the centre cube. A portion of the latter is again enlarged 20 times on the right. The tiny dash within that cube is the distance that would be traversed in 300 years by an ion rocket – one of the more advanced types under development.

Within a year such a vehicle would be moving almost as fast as light (186,000 miles a second). From then on, if the thrust of the engines remained constant, the rate of gain in speed would decrease, approaching zero but not reaching it, in the manner that mathematicians call 'asymptotic'. The engine thrust, no longer able to boost the speed to any degree, would produce strange effects, from the viewpoint of an observer back home. The clock rate – the inherent property 'time' – on board the vehicle would slow down, approaching zero but never quite getting there. The weight of the ship and its occupants would increase; and they would be foreshortened in the direction of flight.

At the mid-point of a journey to one of the nearest sunlike stars, 12 light years away, such a vehicle would be travelling at 99 per cent of light's speed. To land on a planet at its destination it would then have to decelerate at the same rate, with time still dragging its heels, until, during the final year of this outward journey, the speed dropped appreciably below that of light.

These peculiar manifestations would not be evident to those on board. Time would seem normal, as would their shape and weight. Assuming they then returned to earth in the same manner, the time for the entire journey, as measured on earth, would be twenty-eight years (a ray of light would have made the round trip in twenty-four years). However, the passengers would be only ten years older than when they started.

In a sense this is because of the peculiar relationship between time, as a dimension, and the speed of light. When we throw a light switch, the room seems illuminated instantaneously. In the world of our daily lives, the speed of light seems infinite and time seems invariable. But if one could accelerate almost to the speed of light, then strange things would happen to time and the other dimensions. Von Hoerner pointed out that the longer such a journey was, the more extreme the effect, until finally, from the point of view of those back home, time on the space ship would virtually come to a halt. Thus, in a vehicle whose engine accelerated and decelerated it, with a force equal to gravity, a round-trip journey that seemed twenty years long to the voyagers, carrying them to a point 137 light years away, would bring them home to a world 270 years older than when they left it. A sixty-year journey, by clocks on the spacecraft, would bring them back 5,000,000 years after their departure. They would have reached out 2,500,000 light years – further than the nearest galaxy like our own.

Purcell said flatly, in his Brookhaven lecture, that the Theory of Special Relativity, in predicting such strange effects, 'is reliable'. If it were not, he told his atom-smashing audience, 'some expensive machines around here would be in very deep trouble.' Protons in the big Brookhaven accelerator are boosted to 99·948 per cent of the speed of light and their mass increases in the manner predicted by relativity. Those designing the Brookhaven machine had to take this into account or it would not have worked.

Furthermore, the slowing of time, on the atomic level, is demonstrated by the extended lifetimes of the nuclear particles known as muons, when travelling almost at the speed of light. Muons are produced when atoms are shattered by high-energy collisions, such as those in an atom-smasher or the impacts of the high-energy atomic nuclei known as cosmic rays, plunging into the atmosphere from space. So brief is the lifetime of a muon that it decays into other particles almost instantaneously, its average lifetime being 2·2 millionths of a second. Since muons produced by cosmic rays are generated near the top of the atmosphere, they should not survive long enough to reach the surface of the earth despite their high speed. Yet, precisely because of this speed, their lifetimes are prolonged in the manner predicted by Einstein, and muons rain steadily on the earth. In other words, their time, as we see it, is slow. If this happens on the atomic level, would it apply to entire human beings? It is now widely believed that this would be the case. In fact, it is hard to see why it should be otherwise.

Having argued persuasively for the reality of time-slowing, Purcell sought to demolish the idea that anyone, on earth or anywhere else, could ever take advantage of it. He analysed the amount of energy required for the previously described round trip to a star 12 light years away, in which a top velocity of 99 per cent of the speed of light would be reached at the mid-point of both the outward and the return journeys. In terms of fuel weight, the most efficient source of energy within our grasp is the fusion reaction of the hydrogen bomb (where hydrogen isotopes such as tritium and deuterium combine to form helium). Even more efficient is the fusion that makes the sun shine. In this reaction, four hydrogen nuclei ultimately combine, under great pressure, to form one helium nucleus. Because the binding energy required to hold the helium nucleus together is slightly less than that in the original hydrogen nuclei, something is left over after the reaction and emerges as free energy. Though it is less than one per cent of the original mass, the energy released is formidable because of the Einstein equation $E = mc^2$ (the energy equals the converted mass multiplied by the speed of light squared).

To be supremely optimistic, Purcell assumed that this solar fusion process could be used with 100 per cent efficiency, although we have not yet learned to use it at all. In fact, we cannot control the less productive hydrogen bomb process. With solar-type fusion it would require 16 billion tons of hydrogen to accelerate a 10-ton capsule to 99 per cent of the speed of light – or bring it to a halt from that speed. The vehicle must be accelerated at the start, halted at its destination, accelerated for the return and halted again at the end. Since the fuel for these manoeuvres must itself be accelerated, the total requirement is enormous: some 500 billion billion billion billion tons.

But, said Purcell, 'this is no place for timidity, so let us take the ultimate step and switch to the perfect matter-antimatter propellant.' From our present knowledge of nature, there does not seem to be any more efficient way to obtain energy than to combine matter with antimatter. When two such substances meet, they mutually annihilate each other, leaving nothing but a great deal of energy in the form of gamma rays. It is the only known process in which matter (and antimatter) can be converted entirely into energy. The gamma rays which are at the short-wave end of the electromagnetic spectrum, could power a rocket by the equivalent of light pressure.

No one has ever really seen any antimatter. The experimenter can observe in his detector evidence that a particle of antimatter existed there briefly, and some accelerators can store bits of antimatter for limited periods, but sooner or later they meet particles of matter and vanish. If one lists all the particles that can be produced by smashing atoms, there is an 'anti' particle corresponding to every one of them. The antiparticle is a mirror image of its counterpart. If the latter carries a negative charge, like the electron, then its antiparticle is positive. The antiparticle of the electron, for example, is known as the positron. It has the same mass as the electron and the strength of its electric charge is the same, but it is positive instead of negative. Discovered in 1932, it was the first hint that there is an ephemeral world of antimatter.

It seems hard to see how even the most advanced civilization could assemble a mass of antimatter and load it on a rocket, along with an equivalent mass of matter. Again Purcell set aside such problems, but he found that his hypothetical journey would still require 400,000 tons of fuel, equally divided between matter and antimatter.

Then there is the problem of interstellar gas and dust hitting the space ship like hailstones on a car's windshield. Space between the stars is not quite a vacuum. There is about one hydrogen atom per cubic centimetre and there are bits of dust and perhaps other material here and there. If you were travelling at 99 per cent of the speed of light, Purcell said, this material would hit the front of your space ship in the form of radiation as intense, per square yard, as that produced by several hundred atom-smashers. 'So,' he said, with calculated understatement, 'you have a minor shielding problem to get over before you start working on the shielding problem connected with the rocket engine.' Because this engine depends on a massive output of gamma rays for its drive, the problem that it raises is formidable, Purcell said, not so much for the passengers as for the inhabitants of earth. Let us imagine that we are standing at the end of a runway watching a jet airliner take off away from us. If we were too close at the start, the blast from the engines would be scorching. But if this were a rocket powered by gamma rays and headed for a distant star, the 'blast' would be devastating to any life behind it (although, as Carl Sagan later pointed out, the earth's atmosphere would act as a shield against such rays).

Purcell, in his lecture to the Brookhaven scientists, paused before ramming home his conclusion: 'Well, this is preposterous, you are saying. That is exactly my point. It *is* preposterous. And remember, our conclusions are forced on us by the elementary laws of mechanics.' Communication by radio waves is probably the best bet, he said, and he referred to Drake's Project Ozma as imaginative and sound.

Nevertheless, he argued, our society is still not mature enough to engage in a large-scale search:

> We haven't grown up to it [he said]. It is a project which has to be funded by the *century*, not by the fiscal year. Furthermore, it is a project which is very likely to fail *completely*. If you spend a lot of money and go around every ten years and say, 'We haven't heard anything yet,' you can imagine how you make out before a congressional committee. But I think it is not too soon to have the fun of thinking about it, and I think it is a much less childish subject to think about than astronautical space travel. In my view, most of the projects of the space cadets are not really imaginative. . . . All this stuff about travelling around the universe in space suits except for *local* exploration [within the solar system] which I have not discussed – belongs back where it came from, on the cereal box.

Purcell's seeming annihilation of the idea of travel to other solar systems did not discourage everyone. In fact, it stimulated the suggestion that perhaps we may some day be able to propel the entire earth to another part of the galaxy. This was proposed, in somewhat lighthearted fashion, by Darol Froman, who had been Technical Associate Director of the Los Alamos Scientific Laboratory in New Mexico. He was well equipped for the discussion, since the Los Alamos Laboratory, operated for the Atomic Energy Commission by the University of California, is headquarters for American research on reactors for space flight. In a talk to the Division of Plasma Physics of the American Physical Society in November 1961, he noted that the sun eventually will burn out and discussed whether or not, before that dark day, it might be possible to push the earth into another solar system. The energy for this grandiose scheme would be obtained by fusion reactions, using sea-water as the fuel source.

Because the oceanic supply of deuterium, the heavy form of hydrogen used in the hydrogen bomb, is insufficient to push the earth great distances, Froman proposed that it would be more reasonable to use the reaction that occurs in the sun (combining four hydrogen nuclei to form a helium nucleus), even though we are a long way from learning how to do this. The process would make it possible to use hydrogen nuclei that are abundant in the oceans. He suggested that a quarter of this fuel be allocated to escaping the sun's gravity, another quarter be held to manoeuvre the planet into another solar system and the remaining half be used for interstellar propulsion and for light and heat en route. The moon would be forfeited to obtain additional fuel, since, as he put it, the moon 'will be no good to us anyway'. In the absence of sunlight it would be virtually invisible.

Froman's earth-propulsion system could operate for as long as 8 billion years, he said, perhaps enabling a planet to outlive its parent sun and reach solar systems 1,300 light years away. It might even seem preferable, he said, to keep on travelling through the galaxy rather than go into orbit around some other sun. The oceans would then have to be replenished from time to time by gathering water from planets encountered en route. For most of us, Froman said, 'the most comfortable space ship imaginable would be the earth itself. So if we don't like it here because the sun is dying or something, let's go elsewhere, earth and all. We will not have to worry about all the usual hardships of space travel. For example, the radiation problem will disappear because of the atmosphere and because we will be going at low speed. The ease and comfort of this mode of travel are shown in the next slide.'

At this point, there flashed on the screen an idyllic scene showing lady golfers, pine trees and great open spaces.

Actually the problem of propelling large numbers of people to another solar system was discussed as early as 1951 by Lyman Spitzer, head of the Princeton University Observatory. He spoke of a vehicle weighing 10,000 tons, powered by a uranium pile of perhaps 1,000 tons, generating 2 million horsepower of useful energy:

Such a ship [he wrote] could carry thousands of people and vast supplies anywhere in the solar system, and could even navigate to other stars, though many generations would be born, grow up and die on shipboard before such a journey were complete. However, launching such a ship from the Earth's surface to a close circular orbit would be a tremendous undertaking. With the use of chemical fuels, such a launching would require a rocket of some million tons gross weight, an achievement that would seem far, far in the future.

Among those who rebelled against the sober reasoning of Pierce, Purcell and von Hoerner was Freeman J. Dyson at the Institute for Advanced Study in Princeton. It was Dyson who had suggested that a supercivilization might redistribute the material of its solar system to achieve maximum energy from its star. In a letter to the *Scientific American* in 1964 he said the calculations of Purcell and the others were perfectly valid, in so far as they related to fuel requirements for journeys limited to human lifetimes. But what about slower trips? Engines using nuclear power could reasonably be expected to drive large space ships at a speed of a few light years per century, he said. If an intelligent race achieved a life span considerably greater than our own; or if it perfected a method of freezing its citizens, harmlessly, for prolonged hibernation, then journeys reckoned in thousands of years would be conceivable. Thus, he wrote, 'interstellar travel is essentially not a problem in physics or engineering but a problem in biology.'

There is no reason to suppose, he asserted, that others in the galaxy have not solved this problem and that we may not ultimately do so. No doubt many would find thousand-year trips 'unappealing', he said, but added: 'We have no right to impose our tastes on others.'

In another paper Dyson made a startling proposal as to how interstellar vehicles might pick up momentum en route. His scheme was to steal a little bit of the energy with which two very dense stars circle each other. If, he said, a vehicle approached one such star as the star was coming toward it, the gravity of the star would whip the vehicle around in a tight orbit, sending it off into space again with far more energy than it had to begin with. It would be almost as though the vehicle had been hit by a gigantic baseball bat. The star, having transferred to the vehicle some of

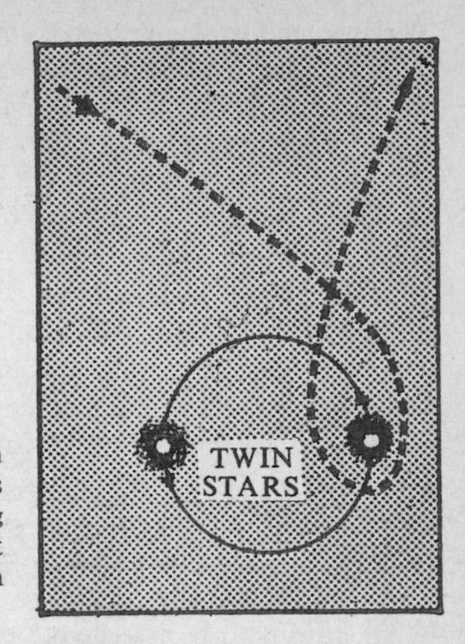

In Dyson's 'gravity machine' an object is fired toward twin stars so that it circles the approaching star and is thrown back by that star's gravity, having gained much additional energy.

the energy with which it was circling its twin, would move a tiny bit closer to the star with which it was waltzing through space.

The most remarkable feature of this procedure was that, even though the vehicle underwent an explosive rate of acceleration – some 10,000 g's – no harm would come to the most delicate passenger or the most sensitive piece of equipment on board. This is because the accelerating force would be applied with almost complete uniformity to every particle of the body or instrument on board. It would not be any more uncomfortable than falling through space.

Thus, said Dyson, a vehicle could very rapidly be speeded up by more than 1,000 miles a second. The best star systems for giving vehicles such enormous accelerations, he said, would be pairs of white dwarfs, tiny 'senile' stars whose density is so great that they may weigh as much as 3,000 tons per cubic inch. 'It may be imagined,' he wrote, 'that a highly developed technological species might use white-dwarf binaries scattered around the galaxy as relay stations for heavy long-distance freight transportation.'

One of the most ardent – and controversial – champions of the feasibility of interstellar travel is Carl Sagan. His main argument was formulated while he was at the University of California in

Berkeley and was presented to the American Rocket Society on 15 November 1962. He sought to show, not only that such travel is possible, but that, if so, 'other civilizations, aeons more advanced than ours, must today be plying the spaces between the stars.'

He argued that radio waves are but a poor way to achieve a meeting of the minds between beings with utterly different histories and ways of thought. Furthermore, the radio does not permit contact between an advanced society and one that is intelligent but not yet in possession of radio technology. Nor does it allow the exchange of artifacts and biological specimens.

'Interstellar space flight sweeps away these difficulties,' Sagan wrote with typical enthusiasm. 'It reopens the arena of action for civilizations where local exploration has been completed; it provides access beyond the planetary frontiers, where the opportunities are limitless.'

He argued that the fuel problem could be overcome by scooping up hydrogen en route and using it to power an 'interstellar ramjet'. The conventional ramjet is used for high-speed vehicles, such as the Bomarc missile, that scoop air into a narrowing duct, thus compressing it. The air is used to burn the fuel, and left-over gases are ejected out the rear. A characteristic feature of the ramjet is that the faster it goes, the more efficient it becomes, since the speed tends to increase the pressure differential between the scooped-up air and the surrounding atmosphere.

The interstellar ramjet was proposed in 1960 by R. W. Bussard who, like Froman, was associated with the Los Alamos Scientific Laboratory. He explained that his engine would be a 'rough analogy' of a conventional ramjet. It would scoop up interstellar material, which is almost entirely hydrogen, using appropriate portions of the hydrogen as fuel for a fusion reactor. The left-over matter would be squirted out the rear, providing an integral part of the propulsion system. Bussard cited the 21-centimetre observations by the Dutch at Leiden revealing the presence in various parts of the galaxy of great clouds of ionized hydrogen that could be collected magnetically. An intake area almost 80 miles in diameter would be required, he said, to achieve the needed velocity for a space ship of about 1,000 tons. 'This is very large by ordinary standards,' said Bussard, 'but then, on any account, interstellar travel is inherently a rather grand undertaking . . .'

Sagan conceded that the problem of scooping up enough hydrogen was staggering. Amplifying Bussard's calculations, he found that in ordinary interstellar space, with only one hydrogen atom per square centimetre, the sweeping system would have to be 2,500 miles in diameter. However, within hydrogen clouds, where the density may reach 1,000 particles per cubic centimetre, the intake could be, as Bussard pointed out, less than 80 miles wide. Perhaps, said Sagan, 'starships' dart from one such cloud to another. Furthermore, he said, there may be some way to ionize material in the non-ionized clouds that predominate in space, so that it can be collected magnetically. Or the starship could select paths through clouds of material that is already ionized. By magnetic techniques it may also be possible to divert particles from the passenger area, thus overcoming the radiation hazard without heavy shielding. The huge vehicle would have to begin its journey, Sagan said, with the aid of fusion-powered rocket stages. The ramjets would be used only when clear of the earth.

In his talk to the Rocket Society, Sagan said his argument was designed to 'lend credence' to the possibility that interstellar vehicles may become feasible for us 'within the next few centuries'. We can expect, he added, 'that if interstellar spaceflight is technically feasible – even though an exceedingly expensive and difficult undertaking, from our point of view – it will be developed.'

He carried the possibilities of time-slowing even further than Purcell and von Hoerner. In continuous acceleration or deceleration (that is, if the engine were kept on), journeys that reached as far into the universe as galaxies millions of light years away would still be possible in the lifetime of the passengers, even though it was questionable whether any civilization would exist on their return. As noted earlier, from the earthly point of view, time on such an extended journey would virtually come to a halt. In fact, Eugen Sänger, head of the Institute of Jet Propulsion Physics at Stuttgart, Germany, had calculated that, with an acceleration no greater than that of the earth's gravity, even the most distant parts of the visible universe could be reached within forty-two years, space-ship time. The incentive, said Sagan, would be greater for journeys to nearby solar systems. Even then, those on the home planet would have to wait perhaps hundreds or thousands of years for the return of their astronauts. It must, therefore, be assumed that a highly advanced society would also be stable over very long periods, preserving the records of previous expeditions and waiting patiently for the return of others. According to this hypothesis civilizations throughout the galaxy probably pool their results and avoid duplication. There may be 'a central galactic information repository' where knowledge is assembled, making it far easier for those with access to such information to guess where, in the galaxy, newly intelligent life is about to appear – a problem very difficult for us, with only our own experience on one planet to go by.

Sagan likened the 'spacefaring' societies to those of the European Renaissance that sent voyagers eagerly in search of new worlds on our own planet. He suggested that such societies might send out expeditions about once a year and, hence, the starships would return at about the same rate, some with negative reports on solar systems visited, some with fresh news from some well-known civilization. 'The wealth, diversity and brilliance of this commerce,' he said with an exuberance reminiscent of Tsiolkovsky, 'the exchange of goods and information, of arguments and artifacts, of concepts and conflicts, must continuously sharpen the curiosity and enhance the vitality of the participating societies.'

On the assumption that there are about a million worlds in the galaxy capable of such feats, Sagan proposed that they would visit one another about once in every thousand years and that scouts may have visited the earth from time to time in the past – perhaps a total of 10,000 times over the full span of the earth's history. One or two million years ago such visitors would have observed the emergence of primates ancestral to man and may have decided to step up the frequency of their visits to once every thousand years, Sagan said.

Is it possible, he asked, that they have visited us since the dawn of civilization? Without making direct reference to the 'flying saucer' episodes of recent years, he dismissed such tales by noting that in the past few centuries, 'when critical scholarship and non-superstitious reasoning have been fairly widespread', there have been no reliable reports of a visitation. However, he urged that myths and legends be re-examined for indications that such may have occurred in the distant past. He cited, for example, a suggestion in the Soviet Union that *The Book of the Secrets of Enoch* may be based on an instance in which a resident of earth was taken home by visitors and then returned to tell his bewildered countrymen about his adventures. The book, also called the *Slavonic Enoch*, is one of those, known as pseudepigrapha, which were purportedly written by Biblical figures but are not accepted as such.

In it, Enoch tells how there appeared to him in a dream

> two men very tall, such as I have never seen on earth. And their faces shone like the sun, and their eyes were like burning lamps; and fire came forth from their lips. Their dress had the appearance of feathers: their feet were purple, their wings were brighter than gold; their hands whiter than snow. They stood at the head of my bed and called me by my name. I awoke from my sleep and saw clearly these men standing in front of me.

They told him they had been sent by God to bring him to heaven. Then, he continued, the two beings 'took me on their wings and placed me on the clouds. And lo! the clouds moved. And again (going) higher I saw the air and (going still) higher I saw the ether, and they placed me in the first heaven.'

In this manner Enoch visited a succession of seven heavens, observing a multitude of wonders, including flying creatures with the feet and tails of lions and the heads of crocodiles. In the seventh heaven he met God and was instructed in the secrets of nature and of man. He was told how the earth was formed and the secrets of the planets and stars. All of this he wrote down in 366 books and, when he returned to earth, sought to impart his wisdom to his fellow men.

Frank Drake points out that there is a somewhat similar episode in the first three chapters of Ezekiel, in the Old Testament, in which the prophet is visited by four winged creatures of strange appearance, accompanied by 'living' wheels, and is miraculously carried to the captive Israelites.

Another ancient legend of the kind that, Sagan argued, should be examined critically appears in the report by Berosus of the manner in which civilization came to Sumeria – the most ancient of all civilized societies. Berosus was a Babylonian priest and historian who lived about 280 B.C. His accounts of the flood and the creation, which we have only at second hand, are very like those of the Old Testament. In the early days, he said, the people of that ancient land between the Tigris and Euphrates Rivers 'lived without rule and order, like the beasts of the field.' Then, he said, there arose from the sea 'an animal endowed with reason, who was called Oannes'. Its body was like that of a fish, but with a second head beneath the fish's head and with feet on its tail like those of a man. This creature, who was able to speak, returned nightly to the ocean, being amphibious.

> This Being, in the day-time [according to the account], used to converse with men; but took no food at that season; and he gave them an insight into letters, and sciences, and every kind of art. He taught them to construct houses, to found temples, to compile laws, and explained to them the principles of geometrical knowledge. He made them distinguish the seeds of the earth, and showed them how to collect fruits. In short, he instructed them in everything which could tend to soften manners and humanize mankind.

Sagan later expressed doubt that either of these accounts was based on a visit from another world, but said the legends of primitive peoples have described, in recognizable form, encounters with a superior civilization. This, he felt, was an incentive for searching the records for evidence of a more exotic visit. He also considered it 'not out of the question' that relics of such visits may be found and suggested that a hidden base may be discovered, perhaps on the far side of the moon, placed there to provide continuity for succeeding expeditions. A remote location would be used, he said, lest the base be destroyed by weathering during the many centuries between visits and to avoid meddling by inhabitants of our planet. When high resolution photographs are made of the moon, as a prelude to landing men there, he said, the possibility of such a base should be kept in mind.

Frank Drake proposed, instead, that such early visitors might have left artifacts for us to find – perhaps as a first step in establishing contact. To preserve such clues from the workings of time and from tampering by primitive inhabitants, they might have been buried in limestone caves, Drake said. Such caves would clearly attract the attention of archaeologists. The artifacts might be tagged with radioactive isotopes whose artificial origin would be evident to any sophisticated investigator. The cache, Drake said, 'would then remain invisible until radiation detectors were developed.'

Drake has calculated that to transmit one pulse – that is, one 'bit' of information – 1,000 light years by radio, would cost only five cents. Hence, few of those who have explored the problem of interstellar communication accept Sagan's argument in favour of travel. In particular, they consider it a grossly uneconomical method of searching for intelligent life, even though it may be the only way to discover societies before they acquire an advanced technology. If beings in other worlds have learned to live long or to hibernate, they may be able to travel between neighbouring solar systems by coasting for a few centuries, instead of constantly accelerating or decelerating, although in so doing they would forfeit the time-slowing effect. But one wonders, despite Sagan's eloquent arguments, if such trips, with all of their cost and discomfort, are necessary.

Biological specimens from another world would be of great interest, but it seems at least possible that they could be replicated by means of radio signals. The human egg cell, the size of a grain of dust, contains a mass of long-chain molecules on which are coded the information needed to construct a human being. The information is voluminous – there are probably billions of 'bits' coded into the egg – and to send it all by radio, particularly at the slow rate that may be necessary for interstellar distances, would be tedious. But by the time we can build interstellar ramjets our communications ability will certainly be much improved. We should be able to tell 'them' how to build a car, a cow, a rose and a man. Nor, if radio were used, would anyone have to worry over the possibility that an arriving space ship might be carrying germs dangerous to the world being visited. Perhaps, in fact, there are galaxy-wide immigration laws forbidding travel in person.

Although even some of the more open-minded scientists shrug off Sagan's arguments, citing the problems and limitations, the question arises: Are we arrogant to believe that we understand these limitations? Are there pertinent phenomena or peculiarities of nature unknown to us, like Philip Morrison's hypothetical 'Q waves'? The reasoning of men like Purcell is based on physical laws that seem immutable. We have no reason to suspect there is anything wrong with them. Yet it is well to call to mind the foresight of Benjamin Franklin, when he wrote to Joseph Priestley, the discoverer of oxygen, in 1780:

> It is impossible to imagine the height to which may be carried, in a thousand years, the power of man over matter. We may perhaps learn to deprive large masses of their gravity, and give them absolute levity, for the sake of easy transport. Agriculture may diminish its labour and double its produce; all disease may by sure means be prevented or cured, not excepting even that of old age, and our lives lengthened at pleasure even beyond the antediluvian standard. O that moral science were in as fair a way of improvement, that men would cease to be wolves to one another, and that human beings would at length learn what they now improperly call humanity!

17

Is There Intelligent Life on Earth?

THE question that forms the title of this chapter was pinned on the door of Frank Drake's office in Green Bank, West Virginia, only half in jest, for it proved the crucial unknown to emerge from the conference privately held there in November 1961.

The conference, whose task was to discuss whether it might be possible to contact other worlds, was organized after the excitement generated by Drake's Project Ozma and the Cocconi-Morrison letter had penetrated the venerable halls of the National Academy of Sciences. Lloyd Berkner, Acting Director of the National Radio Astronomy Observatory in Green Bank, who had given the go-ahead for Ozma, was also chairman of the Academy's Space Science Board. The latter had been formed in 1958, shortly after the launching of the first earth satellites, to set forth national space goals that would be scientifically sound. Its membership included two Nobel laureates, both of whom were deeply interested in problems relating to extraterrestrial life. One was Harold C. Urey, Professor-at-Large of Chemistry at the La Jolla campus of the University of California, whose formidable intellectual energies were being directed toward meteorites, including the perplexing carbonaceous chondrites. The other, Joshua Lederberg, was Professor of Genetics at Stanford University and chairman of the Space Science Board's Panel on Exobiology.

It was not surprising, therefore, that the board decided, in 1961, to call an informal conference at the observatory in Green Bank to assess the possibility of communication with other worlds. J. P. T. Pearman of the board staff did the organizing and, as noted in the first chapter, it was decided not to make any public announcement. Those invited were, to a large extent, the dramatis personae of this book, though not all could come. The ones who did, in addition to Pearman, were:

Otto Struve, director of the observatory, host to the conference and its chairman. His slow-rotating stars had convinced him and many others that solar systems like our own are common.

Dana W. Atchley, Jr, president of Microwave Associates, Inc., specialist in communications technology and donor of the parametric amplifier that was a key component of Project Ozma.

Melvin Calvin, who won the Nobel Prize in chemistry during the conference, and who had spelled out in detail the chemical processes by which life probably evolves from inanimate matter.

Giuseppe Cocconi, who, with Morrison, had proposed a search for signals on the 21-centimetre wavelength.

Frank D. Drake, who independently had initiated a search on 21 centimetres, using the 85-foot antenna at Green Bank.

Su-Shu Huang, who had calculated what types of stars would have habitable zones around them large enough and enduring enough to make them likely abodes for worlds like our own.

John C. Lilly, head of the Communication Research Institute in the Virgin Islands, where he was studying the possibility of communication between man and another rather intelligent species, the dolphin.

Philip Morrison, co-author with Cocconi of the historic proposal for a search on 21 centimetres.

Bernard M. Oliver, Vice President for Research and Development of the Hewlett-Packard Company and one of those who had studied in some detail the problems of interstellar communication.

Carl Sagan, with Calvin a member of the Panel on Exobiology of the Space Science Board and probably the most enthusiastic of the 'exobiologists'. As noted in the previous chapter, he believes that travel between solar systems may be commonplace.

> The purpose of the discussion [Pearman wrote in his semi-official account] was to examine, in the light of present knowledge, the prospects for the existence of other societies in the galaxy with whom communications might be possible; to attempt an estimate of their number; to consider some of the technical problems involved in the establishment of communications; and to examine ways in which our understanding of the problem might be improved.

On the first day, the participants were enthralled as John Lilly told of his research into interspecies communication. Only a few months earlier his book, *Man and Dolphin*, had created somewhat of a sensation and his optimism concerning the prospects of being able to 'talk' with the bottle-nosed dolphins had evoked some outraged reaction in the scientific community.

In short, Lilly was controversial and provocative; but his credentials were good. He had been trained at the California Institute of Technology, Dartmouth, and the University of Pennsylvania, where he was made a Doctor of Medicine and later held twin associate professorships, one in medical physics and the other in experimental neurology. In 1953 he moved to the National Institute of Mental Health as a newly commissioned surgeon in the Public Health Service and became chief of the Section on Cortical Integration in the Institute's Laboratory of Neurophysiology. His scientific respectability was attested by his membership on a number of federal panels, such as the Scientific Advisory Board to the Air Force Office of Scientific Research, and the Scientific Advisory Committee of the Graduate School of the National Institutes of Health. His dolphin research was supported at first by the Navy and the National Institutes of Health and later by the Air Force and NASA.

Lilly had long been interested in the brains of sea-going mammals, in part because of their immense size. The brain of a chimpanzee weighs only about one-quarter that of a human being, whereas the brain of a whale may be as much as six times heavier and that of an elephant is up to four times as heavy as a man's. In large measure this is because the brains of big animals are coarser and they probably have considerably fewer nerve cells than ours. What is remarkable about the brain of the bottle-nosed dolphin, according to Lilly, is not only its size, which is slightly larger than that of an average man, but the fact that, under a microscope, the density of nerve cells seems comparable to that in the human brain. In fact, he said, the cortex or outer layer of the dolphin brain is richer in folds and other structures than the equivalent part of the human brain.

In his book, Lilly sought to document the intelligence of these animals and the complexity of their 'language'. While they can be taught, in captivity, to vocalize outside the water for exhibition purposes, their natural mode of communication is through water and extends into a frequency range far above that of the human ear. They produce a variety of squeaks, whistles, creaks and other sounds, sometimes in very rapid succession. Lilly could easily recognize their distress call – a pair of whistles – and tells of an incident which, to him, suggested the possibility that they have a complex language that can be used to make very specific requests.

The episode involved a dolphin that, during an experiment, apparently had become so chilled that it was unable to swim. Placed back in the main tank with two other dolphins, it sank to the bottom, where it was bound to suffocate unless it could reach the surface to breathe. However, it gave the distress call and the other two immediately lifted its head until the blow-hole was out of water, so that it could take a deep breath. It then sank and a great deal of whistling and twittering took place among the three animals. The two active ones then began swimming past the other so that their dorsal fins swept over its ano-genital region in a manner that caused a reflex contraction of the fluke muscles, much as one can make a dog scratch itself by rubbing the right spot on its flank. The resultant action of the flukes lifted the animal to the surface and the procedure was repeated for several hours until the ailing dolphin had recovered.

He told of other instances in which these animals had come to the assistance of their fellows and of some cases, even, where they reportedly helped people floundering in the water.

'It is probable,' he wrote, 'that their intelligence is comparable to ours, though in a very strange fashion.' We may be faced, he added, 'with a new class of large brain so dissimilar to ours that we cannot within our lifetime possibly understand its mental processes.' Instead of having speech centres as human brains do, the dolphin brain 'may be doing something else entirely than what we do with our brains.' In his day-to-day association with these animals he had observed sufficient humanlike behaviour, he said, to encourage subtly his belief 'that we shall eventually communicate with them'.

Lilly then gave free rein to his imagination in discussing the implication of such an achievement. Dolphins could be used by one government to scout out the submarines of another; they could smuggle atomic bombs into enemy harbours, pick up missile nose cones at sea, help rescue the pilots of downed planes, or serve on underwater demolition teams; all of this being contingent on their loyalty to one side. He recognized the possibility of defection, and realized, too, that they might prove to be pacifists. As a form of psychological warfare, he said, they might be persuaded to sneak up on hostile submarines 'and shout something into the listening gear'. More valuable, he added, would be the gain in scientific knowledge, both from the experience of learning their 'language' and methods of thought and from the information such animals could provide about oceanic life, their own mysterious method of navigation, and so forth.

Lilly argued that to achieve a meeting of the minds with another species on earth – a species less evolved than ourselves, at least in the technological sense – was very like the problem of communication with some higher technology in another world. To prove that communication with a species, such as the bottle-nosed dolphin, is not possible, he said, would take 'a very long time, a lot of research, and the exploration of many possible methods'. The parallel with the search for intelligent life elsewhere was obvious.

It was inevitable that Lilly's listeners should wonder: what if another world were entirely covered with water and this global ocean were dominated by a highly intelligent species, like the dolphins? Should we expect to hear signals from them? It seems unlikely. Huang believes that dry land is one prerequisite for a civilization that can communicate. Life that was purely aquatic could not use fire. As Morrison pointed out, it would be less likely to develop an interest in astronomy, since the creatures would not see the stars except when they poked their heads out of the water, and, without hands, they could not build telescopes. Furthermore, it can be argued that, at least on earth, intelligence appeared in response to the challenges of life on land. The only smart animals of the sea are aquatic mammals whose ancestors presumably were land mammals. The stupidity of fish of comparable size, such as the sharks, is legendary. It would appear that the whales, dolphins and other cetaceans were set on the road to intelligence while living on land.

After Lilly's presentation, Drake, like a true scientist, sought to formulate the conference's central problem as an equation. He wrote it on the blackboard as follows:

$$N = R_{*} f_p \, n_e \, f_l \, f_i \, f_c \cdot L$$

The idea that it expresses is simple enough. The letter N, on the left, represents the number of civilizations in the galaxy that are currently capable of communicating with other solar systems.

All of the expressions on the right side of the equation are factors affecting this number. When multiplied together they give the number that we wish to know – the number of communicative societies. It was agreed that this number is crucial, for if it is large we can expect to find a civilization in our immediate part of the galaxy, whereas if there are only a handful of such civilizations they are probably separated by tens of thousands of light years; to locate them amid the host of 'dead' solar systems would be extremely difficult, and the distance would be so great that only a few messages could be exchanged within a time span comparable to the entire elapsed history of the human race.

The factors appearing on the right side of the equation were as follows:

Factor One (R_{*}): The rate at which stars were being formed in the galaxy during the period when the solar system itself was born.

This would determine the number of stars in the galaxy near which intelligent life may recently have reached maturity – recently, that is, in the sense of the past few hundred millions of years. The astronomers at the meeting said a conservative estimate would be about one new star per year.

Factor Two (f_p): The fraction of stars with planets.

Here the experience of participants like Struve came into play. If, when stars are formed, the left-over material either coalesces into a twin star or a system of planets, then half of all stars have planets, since it can be seen that half are in two-star systems. If, however, in the formation process the left-over material is sometimes disposed of in other ways, such as forming asteroids or being flown off into space, then the fraction of stars with planets might be as low as one fifth.

Factor Three (n_e): The number of planets, per solar system, with an environment suitable for life.

This had been investigated at some length by Su-Shu Huang, as noted in Chapter 7. Although the uncertainties are great, it was proposed that the figure for this galaxy probably lies between one and five.

Factor Four (f_l): The fraction of suitable planets on which life actually appears.

This was Calvin's special province. He and Sagan set forth the argument outlined in Chapter 9, that on such planets, given a time period measured in billions of years, life must sooner or later appear. As Sagan put it in one of his papers, the production of self-replicating systems is a 'forced process' which inevitably occurs 'because of the physics and chemistry of primitive planetary environments'. Hence, the group agreed that this factor was one.

Factor Five (f_i): The fraction of life-bearing planets on which intelligence emerges.

The group was impressed by Lilly's argument that more than one intelligent species had evolved on this planet, although the example of the dolphins also raised the possibility of life that was intelligent but would not become technological.

The most telling argument here seems to be the characteristic pressure of life that drives it into every nook and cranny of the environment where, by some marvel of adaptation or ingenuity, sustenance can be found. As we look about us on the earth we see examples of this on every hand. If there is a way to live by swimming, fins will evolve. If there is a way to live by walking, legs will appear. If there is a way of life in the sky, some animals will develop wings. Mites have found a way to survive on peaks near the South Pole. Algae live in the scalding water of hot springs. In the perpetual night of the oceanic trenches, 6 miles below the waves, or adrift in the high atmosphere, one finds life. The wonderful process of evolution has, over the billions of years, pushed life into every 'ecological niche' that one can imagine. As in the free enterprise system, if there is an odd way to make a living, someone will discover it sooner or later and prosper. So far as we can see, the most successful way to live on earth is to manipulate the environment by what we call intelligence. In its early stages this involves the use of tools, plants and domesticated animals to obtain clothing, shelter and food. Eventually it leads to almost complete transformation of the landscape, as is evident today to any airline passenger on our own planet.

The fork in the road that led to intelligence began very far back in the history of life on earth, for even such lowly creatures as the insects have tiny, compact computers no larger than a pinhead that are primitive brains. At least one biologist and one paleontologist – Blum at Princeton and Simpson at Harvard – doubt that intelligence is an inevitable fruit of evolution. In fact, they believe that if there are any intelligent beings in this galaxy, they are extremely distant. But for a long succession of evolutionary 'accidents', according to Simpson, we would be no smarter than a cabbage plant.

In his book *This View of Life*, he concedes that intelligence is 'a marvellous adaptation' which has 'survival value in a wide range of environmental conditions'. It would therefore be favoured by natural selection on a variety of planets, once it appeared. However, he says the factors that produced a thinking creature on earth 'have been so extremely special, so very long continued, so incredibly intricate that I have been able hardly to hint at them here. Indeed, they are far from all being known, and everything we learn seems to make them even more appallingly unique.'

His book is an outcry against the search for life beyond the earth, and yet he draws attention to one of the problems that those more enthusiastic about space exploration hope can be enlightened by such activity. This concerns the extent to which the paths of evolution on earth have been typical of life elsewhere. Even if life on Mars is now extinct, study of its remains, if any, should help throw light on this question.

Simpson's book had not appeared at the time of the Green Bank conference. Of those at that meeting, the one who had probably thought the most about this problem was Philip Morrison, a nuclear physicist by trade, but a voracious reader in a wide variety of fields. He argued that intelligence would always appear, sooner or later, because of 'convergence'. This is the tendency of species, evolving along highly diverse routes, to converge towards life forms that, because of certain basic laws, resemble one another. He cited, for example, three fundamentally different kinds of animals that evolved into the same shape: one a reptile, one a fish and the third a mammal. The reptile was the plesiosaurus, which became extinct at the end of the Mesozoic Era, some 100 million years ago. It appears to have been a terror of the deep in its heyday. The fish that he cited was the tuna, and the mammal was the dolphin. Although these creatures evolved from ancestors utterly different from one another, their final paths of evolution were determined by the laws of hydrodynamics. There is an ideal shape for bodies six feet in length that wish to swim. By the slow, cruel process of mutation and survival of the fittest, each of these branches of the animal kingdom produced a creature that conformed closely to this ideal. Another instance of conversion would be the appearance of eyes as the fruits of three separate lines of evolution: in vertebrates, insects and molluscs.

So far as intelligence is concerned, Morrison saw an example of convergence in the appearance, on earth, of two rather intelligent kinds of life: men and the cetaceans (including dolphins). Being mammals, both have a common ancestor that was furry and suckled its young, but it was certainly of very low intelligence. Yet the opportunities that the environment offers to intelligence caused this characteristic to emerge at the end of two quite separate lines of evolution. An added peculiarity of man's evolution, Morrison noted, is that he has eliminated all neighbouring species; he stands alone, with no close relatives in the family tree of the primates. Whether the competitors for this ecological niche eliminated their rivals by clubbing them with bats, by collecting food more efficiently or by dominance and intermarriage is unknown. This took place so long ago that there is not even any clear fossil evidence of what occurred. The fact, however, that this happened in earth suggests that intelligent land animals like man do not tolerate close competitors.

The conference at Green Bank decided that a factor of one should be assigned to the emergence of intelligence – meaning that it would arise, eventually, on virtually any planet where there is life.

Factor Six (f_c): The fraction of intelligent societies that develop the ability and desire to communicate with other worlds.

Here the conferees felt the need of a sociologist, anthropologist or historian, although, as Morrison pointed out in one of his lectures, even the specialists in these fields lack the knowledge needed to be of much help. 'We are trembling on the edges of speculation which our science is inadequate to handle,' he said, for we have no adequate theory of the social behaviour of complex societies. 'Our experience, our history, is not yet rich enough to allow sound generalization.' How likely is it, he asked, that populations of men, 'or manlike things', would evolve a technology with explosive speed, as we have? He pleaded with the historians and sociologists to seek out guiding principles.

At Green Bank, however, Morrison did argue that the principle of convergence would apply. Specifically he sought to show that the early civilizations of China, America and the Middle East evolved separately but nevertheless along similar lines. Thus, when the Spaniards reached Central and South America, they found a civilization that was several thousand years behind Eurasia in that it had not developed an alphabet. In fact, the use of paper was unknown. Yet its monuments show that symbolic pictographs were employed to record historical events. Had the Incas and Aztecs been left to their own devices, Morrison argued, they would soon have progressed to more stylized forms, like the hieroglyphics of ancient Egypt or the early Chinese ideographs.

His point was that civilization, including primitive writing, evolved independently in Eurasia and America. The continued isolation of those regions from one another during ancient times is seemingly borne out by the fact that the contagious diseases of one were unknown in the other.

In a later discussion he acknowledged that our technology could not have come about but for that remarkable event in human history, the Renaissance. In this case there was no convergence. There was only one Renaissance and it took place in only one culture. Was this, then, a freak occurrence that would not be likely to occur in other worlds? He cited, for example, the thousand-year history of classical China that evolved gunpowder, maps, printing, the compass and paper, yet levelled into a plateau of stability that endured until the 'new philosophy' in Europe lighted a fire that is still spreading around the world.

Just why this happened in Europe and not in China is still a favourite subject of doctoral theses. Morrison argued that, had the Renaissance not come to Europe, something comparable would have been forced upon China when the status quo had run its course – when the dams had silted up, the land become impoverished, or the rice become diseased. In a different context, such a revolution has, in fact, been taking place in China over the past century.

To be weighed against this optimistic argument was the long record of civilizations that have flowered and died without ever discovering the scientific method and becoming technological. How certain can we be that this discovery is inevitable? Furthermore, it was pointed out, intelligent life may have arisen on planets deficient in the metals on which our technology is based.

As to whether or not a highly advanced society would be interested in making contact, there seemed several possibilities. If a society has conquered poverty, disease, hunger and overpopulation, greatly increased the life span of its citizens, minimizing their labour and their worries, what would be their state of mind? Would they become bored and lazy, losing interest in science, forfeiting their intellectual curiosity? Sebastian von Hoerner, in an analysis published in *Science* shortly after the conference, pointed out that science and technology have been advanced, in large measure, though not entirely, 'by the fight for supremacy and by the desire for an easy life'. Both these forces, he added, 'tend to destroy if they are not controlled in time: the first one leads to total destruction and the second one leads to biological or mental degeneration.' Does this mean that mankind will end up vegetating in front of television, as Bracewell has suggested, or live in automated hermitages, as depicted in E. M. Forster's frightening short story 'The Machine Stops'?

Morrison, in a lecture to the Philosophical Society of Washington, was more hopeful. He proposed that advanced societies throughout the galaxy are in contact with one another, such contact being one of their chief interests. They have already probed the life histories of the stars and others of nature's secrets, he said. The only novelty left would be to delve into the experience of others. 'What are the novels?' he asked. 'What are the art histories? What are the anthropological problems of those distant stars? That is the kind of material that these remote philosophers have been chewing over for a long time.'

Of course, if such societies are numerous, they may be sated with such novelties; but he still argued that they would be interested. To discover a new realm of life would be a coup for such a civilization. If intelligent life is scarce, the motivation to make contact will be far greater – and doing so will be far more difficult. In any case the question is beset with imponderables, such as the motivation of 'people' who have been operating technical institutions for long periods of time.

How universal is curiosity? How persistent is it? From our experience on earth, it would appear that this quality is an essential ingredient of successful intelligence. We see it, in ways that delight us, in our small mammalian friends as they go sniffing about. We know that it is the kernel of science. Morrison proposed that intelligent beings in other worlds may, over great time spans, evolve into species that lack curiosity. 'They turn into some other animal and they're not interested any more,' he suggested.

However, if intellectual curiosity dies, it would seem that intelligence is likely to do so, as well. The ringing words of the Norwegian explorer Fridtjof Nansen come to mind: 'The history of the human race is a continuous struggle from darkness toward light. It is therefore of no purpose to discuss the use of knowledge – man wants to know and when he ceases to do so he is no longer man.'

Taking all these considerations into account, the Green Bank conferees estimated that from one-tenth to one-fifth of intelligent species would engage in attempts to signal to other parts of the galaxy.

Factor Seven (L): Longevity of each technology in the communicative state.

This proved to be the factor most difficult to estimate and most critical in assessing our chances of making contact. It was here that the question arose: Is there 'intelligent' life on earth? Are we smart enough to suppress our aggressions and prejudices to survive the crises that confront us? If we lack that ability, the chances are increased that other civilizations will lack it, too. As the conferees sat around their table in November 1961, they were very aware that their country and the Soviet Union were, in effect, holding nuclear pistols at each other's heads. It was not inconceivable that the typical lifetime of a technology sufficiently advanced to destroy itself is only a few decades. If that is the case, no one is calling us.

We know from history that the civilizations that have arisen in the past on our own planet invariably have fallen. Is ours an exception? This time will it be a nuclear holocaust that brings down all the mighty towers of culture, instead of a new, vigorous society overrunning a senile one?

Morrison noted that on a recent visit to Cornell, Fred Hoyle, Plumian Professor of Astronomy and Experimental Philosophy at Cambridge University, had discussed the fate of highly evolved technologies. He elaborated on the idea in a lecture that he gave at the University of Hull in 1963 entitled 'A Contradiction in the Argument of Malthus'.

His reference was to the theory advanced by Thomas Robert Malthus, an English clergyman and political economist who, at the end of the eighteenth century, proposed that population tends to increase faster than food supply, and that therefore poverty and war inevitably become the instruments for holding the population in check. Hoyle pointed out that this theory influenced Charles R. Darwin in his thoughts on the mechanics of evolution and that Darwin's grandson, Sir Charles G. Darwin, was recently led by it to predict, in his book *The Next Million Years*, that starvation will ultimately be the grim brake on our population explosion. Mankind today, Darwin said, is at its peak; the future will be an increasingly bitter struggle for food and survival.

Hoyle argued that such a Malthusian theory is not applicable to a society like our own, dependent for its life not on a simple agricultural economy like that of the eighteenth century, but on a system of production and distribution that becomes ever more complex as the population grows.

Hoyle pointed out that as long as the number of births exceeds the deaths, the population will grow according to a simple mathematical law – that of the exponential. This is true no matter how small the margin of births over deaths. The effect of the margin is simply to determine the steepness of the curve of population growth. If the present rate continues, Hoyle said, within 5,000 years the total mass of humanity will exceed the mass of all planets, stars and galaxies visible with the 200-inch telescope on Mount Palomar.

Obviously, before this occurs either the birth rate must drop to the level of the death rate or the death rate must increase to match or exceed the birth rate. There are many examples from history that the latter can occur, but no precedents for the former, Hoyle said. 'There is one respect in which the human species has shown not the slightest originality – its excessive reproductive vigour.'

His argument was that a collapse of organization, rather than starvation, was likely to be the limiting factor. It is typical of 'organizations' of great complexity that they are subject to collapse. He cited the behaviour of ice. It holds together until, at a certain temperature, the increased thermal motions of its molecules bring about its collapse into water. Melting is common to a great many substances, he said, and added that such collapses seem to be a common property 'of all organizations, irrespective of whether the individuals happen to be humans or chemical molecules'.

Human society is bound to become much more complex to feed and care for itself, he said, and the more complex it becomes the greater will the disaster be. It could be the consequence of nuclear war: 'Methods of warfare are certain to become more efficient, and it can hardly be doubted that total war a century or two hence would be capable of producing just such a discontinuous change as must arise.' However, he argued that war is not the only way in which the organization of society could collapse.

Another possibility would be the appearance of a germ strain against which adequate defences could not be mustered before it spread epidemic around an overcrowded globe. In any case he doubted that destruction of the human race would be total. 'The tattered remnants of humanity will slowly reform and re-establish themselves.' The struggle for survival would bring into play laws of evolution that have been in abeyance, so far as the human species is concerned. 'If I myself were a survivor,' Hoyle said, peering at his audience through horned rims, 'I would live in constant terror of breaking my glasses. Such a simple event would almost certainly make further survival impossible for me.' The chaos following collapse would thus eliminate evolutionary regressions, such as poor eyesight, that had appeared during the past few centuries.

The chaos would also, be argued, bring forward those with intellectual gifts, for he assumed that libraries or other reserves of information would survive. It would be those who could read, and understand what they read, that would be most fitted to survive. They would have to be ingenious, for they would find a world in which ready raw materials had been exhausted – quite unlike those who built the first industrial economy amidst an abundance of iron, coal, oil and the like. To survive in such a situation, the world's remaining inhabitants would have to work closely together. 'It will be the unco-operative elements that will be rejected. Such individuals prefer to exist in chaos and will finally be submerged.' All of this would make for the selective emergence of a more intelligent and 'sociable' race. Hoyle envisaged that the cycle of catastrophe and recovery would repeat itself. In fact, he spoke of as many as twenty cycles, spread over the next 5,000 years. He showed the successive slow rises and steep falls as a sawtooth pattern, with the intelligence and sociability of the race increasing each time. Such improvements may ultimately enable men to do what Hoyle felt present-day man could not achieve – suppress the birth rate. In his lecture at Cornell he suggested that the break in the sawtooth pattern might finally come about through the establishment of contact with a more advanced society – one that had already achieved stability.

This was what von Hoerner called the 'feedback' effect. He had worked with Frank Drake at Green Bank and his line of reasoning was very like that of participants in the conference. He sought in particular to assess the question of longevity and, like Hoyle, discussed, in his *Science* article, the possibility that a succession of civilizations might emerge on one planet during its long history. However, he did not envisage the original species reviving each time with greater intelligence and sociability. Instead he saw the dominant species wiped out by each cataclysm, with a new civilization emerging 'out of the unaffected lower forms of life'. This would be a far slower process than that envisaged by Hoyle. Instead of 5,000 years, it would probably require hundreds of millions. However, since the first technological society on our planet has emerged roughly midway in the lifetime of the sun as a stable star, there should be plenty of time left for such cycles to rise and fall.

This was but one of a number of possible fates for a civilization suggested by von Hoerner. Others included complete destruction of all life, 'physical or mental degeneration and decay', loss of interest in science and technology, or an unlimited lifetime, though he quickly dismissed that last alternative. If the average longevity is 6,500 years and each habitable planet, in its full lifetime, produces an average of four successive technologies, then, he calculated, only one in every three million stars, at present, has a technical civilization, and the mean distance between such worlds is about 1,000 light years.

This was considerably more pessimistic than the estimate of Sagan, who suggested that some civilizations may achieve a global society before they discover weapons of mass destruction. In any case, he said, if they were able to live with such knowledge for a century, they would be likely to survive as long as their sun remained stable. Sagan doubted that such a civilization would succumb to a natural disaster, such as an epidemic, climate change or geologic upheaval. He estimated that one in 100 thousand stars have advanced societies in orbit around them.

Von Hoerner's calculation of one in three million made it seem impracticable to look for signals from selected stars. The entire sky would have to be scanned continuously and one would not expect other civilizations to beam signals at our sun. Rather, they would use beacons radiating in all directions. It was this idea that was behind the suggestion of Iosif Shklovsky that the Soviet Union scan the Great Spiral Nebula in Andromeda for such a beacon.

In any case, von Hoerner argued, once a civilization made contact with another world, its own life expectancy would be greatly increased; for, through the knowledge that others have weathered the crisis and, perhaps, with guidance as to how this was done, the new member of the galactic community would be able to solve its problems better. On the other hand, if civilizations, once they have reached our level, survive only briefly, few, if any, will live long enough to make contact. Hence, von Hoerner's 'feedback effect' will be negligible. He calculated that the critical longevity is in the neighbourhood of 4,500 years. If civilizations endure a long time, compared to that figure, there is a strong chance that they will establish communications with sister societies, acquiring greatly increased wisdom and therefore an even longer life span. If the lifetimes are short, the likelihood of contact will be small.

In fact, as Morrison pointed out, if our civilization survives only ten years after reaching its present potential for self-destruction, and if this is typical, then, on the average, there is only one communicative site in the galaxy at any one time. There have been others in the past and there will be others in the future, but they do not last long enough to coincide. 'It is a very gloomy prospect, but a possible one,' he said. Indeed, he added, the discovery that this is not true would be one of the greatest rewards of detecting signals from someone else.

Von Hoerner recognized one other possibility, which was that, while the average life span may be short, exceptions may alter the situation fundamentally. If two of the more stable societies have made contact 'once somewhere', he said, it may have initiated a chain reaction, with new civilizations being located and 'saved' before they destroyed themselves. Those who have tasted the excitement of speaking to another world might easily be inspired to long and patient efforts to broaden the network of cosmic wisdom.

The feedback effect was discussed at the Green Bank meeting and, as in von Hoerner's analysis, it was concluded that the mean lifetimes must either be very long or very short. An in-between situation seemed unlikely. Thus it was proposed that the figure is either less than 1,000 years or more than 100 million years. 'Fears that the value of L [lifetime] on earth may be quite short are not groundless,' Pearman wrote in his account of the meeting. 'However, there is at least the possibility that a resolution of national conflicts would open the way for the continued development of civilization for periods of time commensurate with stellar lifetimes.'

If, in fact, the average lifetimes are under 1,000 years, then the mean distance between such civilizations is very large – some thousands of light years – whereas, if the longevity is more like 100 million years, the separation distance is from 10 to a few hundred light years.

For every light year gained in observing range, the number of accessible stars increases rapidly, out past 1,000 light years. Then the increase begins to drop off in some directions because the range reaches beyond the disc of the galaxy. However, this is compensated to some extent by the fact that, toward the core of the galaxy, the star density becomes greater than it is near our sun. In our vicinity the mean distance between stars is nine light years, whereas, toward the centre of the galaxy, it is only one light year. C. M. Cade in Britain has suggested that networks of worlds in communication with one another are more probable in this region of greater star density than in our location in the outer part of the galaxy.

The Green Bank conferees also reviewed 'logical' ways to communicate. They discussed the possibility that some obvious wavelength, such as 21 centimetres, might be used only to attract attention. The signal would contain instructions as to some other, more efficient channel of communications. Morrison has drawn an analogy with the problem of establishing contact with primitive natives. One does not try calling them with television, he said. One sends a man ashore 'who beats a drum'. The more sophisticated channels can come later.

The participants set forth their various ideas on attention-getting and eavesdropping methods, as described in earlier chapters. Drake noted that the use of space vehicles for sending and receiving radio signals would not greatly alter the optimum frequency range. In space, he said, the range would lie between 1,000 and 30,000 megacycles, whereas on earth, because of atmospheric effects, it would be from 1,000 to 10,000.

Huang, Drake and others warned that searching for life in other worlds for hundreds or thousands of years would not hold its excitement for very long. This would apply to both individuals and governments. Drake noted his experience with Project Ozma. He pointed out that constantly acquiring nothing but negative results is most discouraging. A scientist must have some successes or his interest flags. Hence he proposed that a project for detection of extraterrestrial life also carry out conventional research. Perhaps time should be divided about equally between the two, he said.

Morrison injected a further note. He said that to keep the watchers from getting bored it might be wise to 'sneak in' false signals once in a while. It might also be possible, he said, to allocate listening tasks to a number of telescopes, thus reducing the burdern on any one observatory.

It may also be, as pointed out by Huang, that distant worlds have found it impractical to scan all candidate stars and also to beam signals toward them over enormous periods of time. If so, an eavesdropping scheme like the one proposed by Drake would be the only practicable method. Drake believed that, for a reasonable chance of success, such a programme must have the following features:

1. It should simultaneously monitor at least 1,000 channels across the entire breadth of the optimum radio spectrum.

2. The receiving radio telescope should have an antenna at least 300 feet in diameter.

3. A 'very large' computer should be used to store and correlate the received data, eavesdropping by the radar cross-correlation method that he had proposed.

4. To scan enough stars for 'a high probability of success', the system should be operated about thirty years.

Drake estimated the capital investment in such a project at $15 million. (In another analysis Morris Handelsman of the Advanced Military Systems division of the Radio Corporation of America has estimated the cost of a system to scan 200,000 stars within 1,000 light years at $3 billion.) Drake concluded that 'the scale of the undertaking, the shortage of people who are qualified to participate and interested in doing so, and the novel nature of the experiment all militate against early commencement of the project.' Similar pessimism was expressed by Otto Struve. Despite the interest of 'many distinguished scientists', he said, 'those responsible for spending government money on scientific research are necessarily sceptical about financing so expensive a project that does not promise quick results.' He thought a full-scale search in the near future 'unlikely'.

What was needed first, the Green Bank group agreed, was more knowledge concerning phenomena already within reach: the number of nearby stars that have solar systems and the possible existence of life that has evolved elsewhere – specifically on Mars. Also required, as set forth by Pearman in his report, was 'more profound analysis' of the evolutionary and sociological factors that bear on the question.

If it is found that planets are abundant and that life exists even in so grim an environment as that of Mars, then the argument in favour of a search will be enormously strengthened. On-the-spot efforts to see if there is life on Mars are at hand, and Drake urged that renewed efforts be made to find evidence of planets. He pointed out that the cross-correlation methods so useful in picking up weak radar signals would not only help in eavesdropping on distant radio traffic, but could also be applied to the search for planets.

As described in Chapter 6, it had already been possible to detect large planets by looking for 'wiggles' in the motion of their parent stars, but the telltale motions are so slight that they are often hidden amidst the observational errors. Drake suggested that cross-correlation techniques be used to process large quantities of positional data and seek out evidence of wiggles from the inherent errors. As in the case of his eavesdropping proposal, the method was comparable to superimposing radar echoes to raise them above the background 'noise', except that, whereas the code of returning radar pulses is known, the wiggle rhythm reflecting the orbital period of a postulated planet is not. This simply meant that, by computer analysis, all possible periods had to be tested on the data. Drake later applied this method to Epsilon Eridani, one of the targets of Project Ozma, and found indications of a planet six times as large as Jupiter.

Suddenly the seemingly glamourless branch of science known as positional astronomy has taken on new meaning. For generations those who practised it had taken photographs, made careful measurements and stored them away for comparison by future generations. This slow accumulation of data has, as Morrison put it, been viewed as old-fashioned: 'Nobody helps them out; nobody gives them computing machines; nobody has lots of people working on that.'

Hopefully, the situation is changing. Astronomers like Drake look forward eagerly to the accumulation of highly precise data from the new 61-inch astrometric telescope set up by the Naval Observatory near Flagstaff, Arizona. Its chief task is to acquire position information on the nearer stars, particularly the dimmer ones, to determine their distances through triangulation. However, the instrument should also make it possible to see the wave-like motion of stars that have planets.

While we may thus learn the plenitude of solar systems and, by Mars landings, may find that life, also, is plentiful, there is no immediate way to determine the chief unknown in the Green Bank analysis – that of the longevity of a technological civilization. The other factors in Drake's equation can be assessed by experiment or observation, but not that one. In fact, he has suggested that the only way to learn the typical longevity may be by making contact with other worlds. Yet our confidence in the future will be greatly increased if we can survive the next few decades. The life span of a society necessary for a meaningful exchange of ideas with other worlds is one for which there is no precedent on earth. Our civilization would have to endure for thousands of years.

But neither do we have any precedents for the predicament in which we find ourselves. As with von Hoerner's feedback effect, it is an all-or-nothing situation. Either we fail to meet the crisis or we achieve a global society, immune to most, if not all, the factors that have brought down civilizations in the past.

18

Celestial Syntax

NOT long after the Green Bank conference, Frank Drake sent to all the participants a strange communication. It was, he believed, of the sort that might be received as the initial message from another race of beings and it manifested the thought that he and others had given to the question: how can we exchange ideas with those who differ from us in ways we cannot guess – certainly in physiognomy, possibly in methods of thought and logic?

Drake's message consisted of a series of pulses sent in a fixed rhythm, but with many gaps. The pulse he wrote as ones; the gaps as zeros. In other words, it was a binary code, the simplest of all communication systems in that it makes use of only two symbols. It is the one commonly used in computers. His colleagues worked over it, but despite some familiarity with the workings of Drake's mind, they found it so compact that its full meaning was difficult to extract. However, one of them, Bernard Oliver, expanded – and thus simplified – the message. It appears on the next page.

A peculiar feature of this message is that it consists of 1,271 ones and zeroes – in computer terminology, 1,271 'bits' of information. To anyone mathematically inclined, the number 1,271 will be recognized as the product of two prime numbers: 31 and 41. This suggests the possibility that the message should be written out in 41 lines of 31 bits each, with the zeros, since they were pauses in the transmission, left as blank spaces. As shown on page 292, this arrangement is unenlightening.

The other possibility, 31 lines of 41 bits, is shown on page 293.

It is clear that the transmitting planet is inhabited by two-legged creatures that apparently are bisexual and mammalian. The circle at the upper left is their sun, with its planets spaced below it. The man-like creature points to the fourth planet, where the civilization resides. The wavy line emanating from the third planet shows it to be covered with water, and below it is a fish-like symbol. Apparently the civilization has visited this planet and

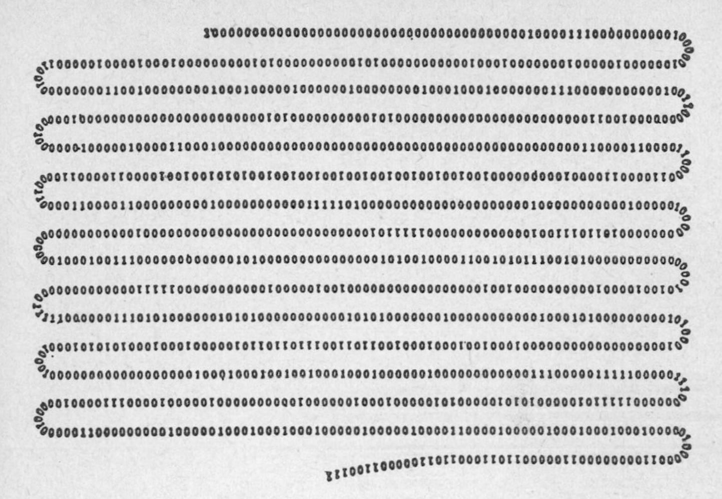

The message of 1271 'bits' (all ones and zeros) as it would appear transcribed on continuous tape.

found it to have marine life. The planets are numbered, down the left side, in a binary code. Symbols across the top represent the atoms of hydrogen, carbon and oxygen, indicating that the chemical basis of life there is like our own. The scale on the right, labelled in the binary code, shows that the creatures are eleven units high, the units presumably being the wavelength of the signal – 21 centimetres. This works out to 7½ feet.

Drake's cryptogram was devised not merely for his amusement and that of his colleagues. He sought to stimulate some hard-headed thinking about the decipherment problem. What was needed, Morrison said, was a new speciality: 'anti-cryptography', or the designing of codes as easy as possible to decipher. He devised a scheme for coaching distant creatures in the establishment of a television link. Such, it is widely thought, would be of immense importance, for both sides would have almost nothing in common. All that we can count on is that all beings sufficiently intelligent for interstellar communication must have a mathematics based on numbers and on the simple concepts of addition, subtraction, division, equality, and so forth. They must be

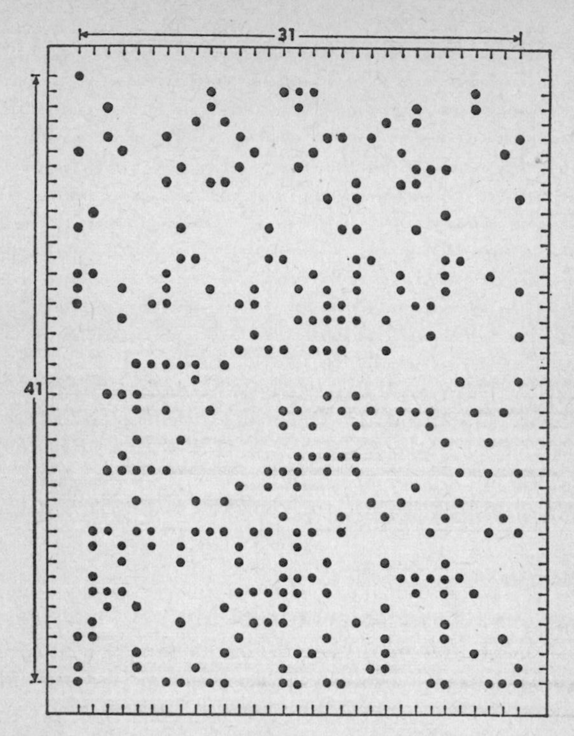

The message arranged in 41 lines.

aware of the structure of the various atoms. Yet their higher mathematics, their logic, their way of representing atomic structure, may differ radically from our own.

Despite these handicaps, it is very probable that they can see. The sense of sight is so important that, as Bernard Oliver pointed out, it has evolved independently in widely diverse creatures on our own planet: flies, scallops and men, for example. Hence the emphasis on establishing television contact – using what Bracewell described as 'sophisticated sign language'.

Morrison's method for making such contact was an elaboration of the 'radioglyph' scheme advanced by the British mathematician Lancelot Hogben in a 1952 lecture to the British

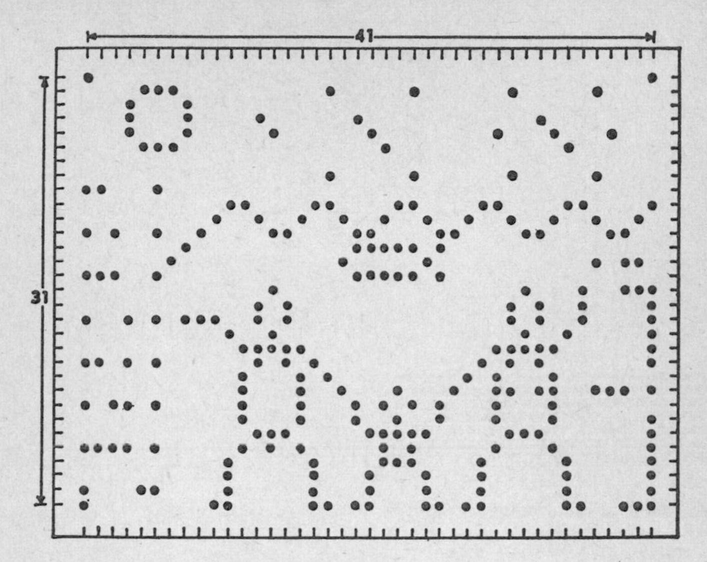

The message arranged in 31 lines.

Interplanetary Society entitled 'Astraglossa, or First Steps in Celestial Syntax'. Hogben had proposed that numbers could be represented as ordinary pulses. Five such pulses would indicate the numeral 5. Mathematical concepts, such as 'plus', 'minus' and 'equal', would each be symbolized by a distinctive signal of some sort – a radioglyph. In Morrison's version, numbers were indicated by square-shaped pulses, whereas 'plus', 'minus', 'the reciprocal of' and so forth were represented by other pulse shapes. Distant beings would be introduced to the meaning of these symbols in terms of simple arithmetic, as follows:

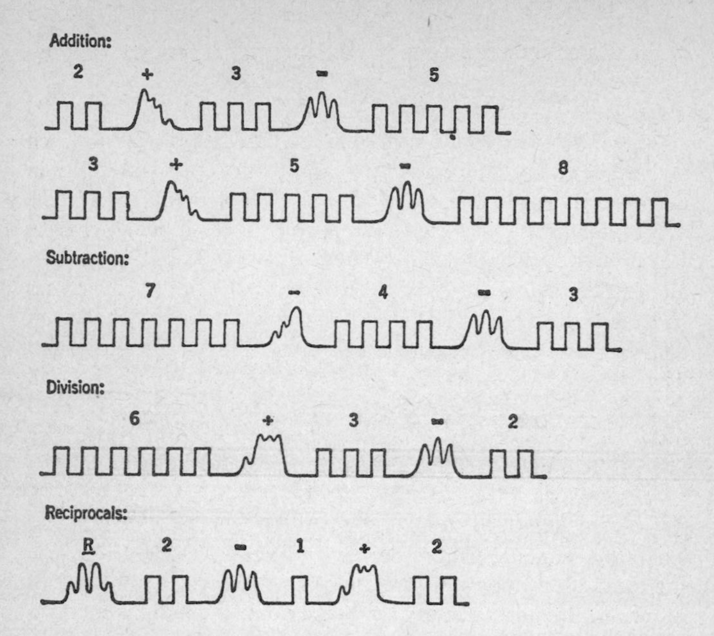

The next problem would be to introduce the concept of pi, the ratio of the circumference of a circle to its diameter. Pi is an 'irrational' number that cannot directly and precisely be expressed in numerals. However, it can be suggested by adding together an indefinite series of fractions whose sum 'converges' on the value of pi. This was how Morrison proposed to do it.

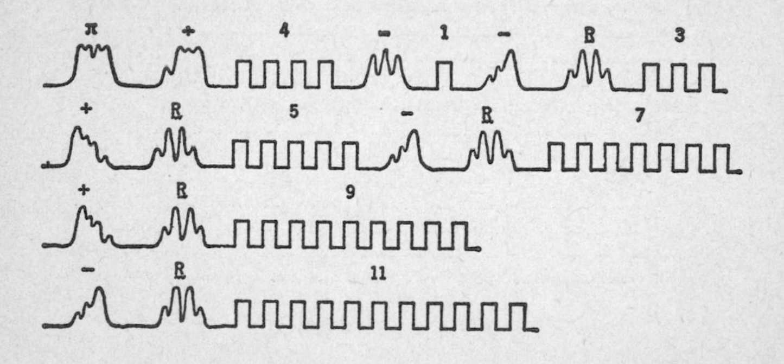

Translated into conventional arithmetic, this means:

$$\pi/4 = 1 - \tfrac{1}{3} + \tfrac{1}{5} - \tfrac{1}{7} + \tfrac{1}{9} - \tfrac{1}{11} \ldots \text{etc.}$$

He would then send a series of extended radio signals, each marked at its start and finish by a pulse, with occasional pulses in between. He would intersperse these with the symbol for pi in the expectation that those on the receiving end would guess what he was driving at; namely, that the lines of radio emission should be aligned, one below the other, so that the seemingly random pulses form a circle (below). This would establish the 'raster' of a television screen – the pattern of its horizontal scanning lines. The next step, Morrison suggested, might be to transmit an illustration of the Pythagorean theorem – reminiscent of the proposals of a century earlier for interplanetary attention-getting (page 296).

The 525 lines of American television channels are scanned 30 times a second, which is fast enough so that our eyes see only a smoothly changing picture. However, there are severe limits on the speed with which a picture can be transmitted over interstellar distances. The farther away the transmitter, the more prolonged must each pulse, or 'bit' of information, be to stand out above the background noise. This already has presented problems in designing the television systems for exploration of the solar

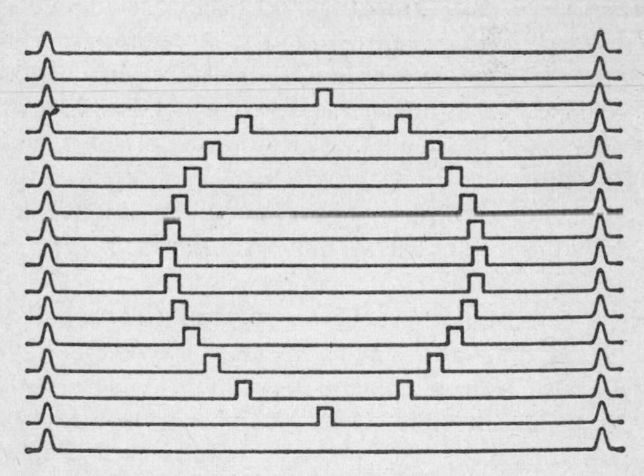

Establishing the TV 'raster'.

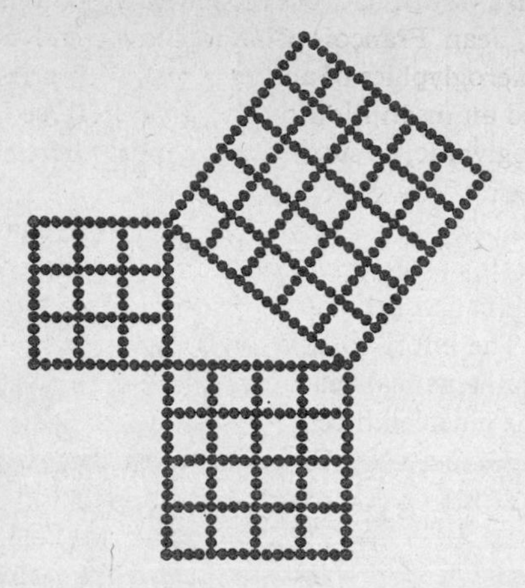

The Pythagorean Theorem on TV.

system – notably the one in Mariner IV that transmitted pictures of Mars 140 million miles to earth. Whereas commercial television stations can send a picture in one-thirtieth of a second, it took Mariner 8½ hours to do so. Each image was reconstructed by computer from 240,000 bits that had been transmitted at a rate of 8·3 per second. While the transmitters used for communication between solar systems would be far more powerful than those of Mariner IV, the distances would be vastly greater. It therefore appears that interstellar sending speeds will have to be far slower, but this should not be too troublesome, since the messages will in any case take years to reach their destination. Also we, in our ignorance, cannot set limits on the skill and ingenuity of our distant friends in devising ways to speed up their sending rates.

Without television it would be very difficult to arrive at a mutually understandable language. Russell F. W. Smith, linguist and Associate Dean of General Education at New York University, has cited the difficulties in deciphering lost languages on our own planet. At the 1963 convention of the Audio Engineering Society in New York, he pointed out that, but for the discovery by Napoleon's troops of a basalt slab at Rosetta near the mouth of the Nile, Jean François Champollion could not have deciphered the hieroglyphics of ancient Egypt. The priests of Ptolemy V had placed an inscription on the stone in three alphabets: the ancient hieroglyphic, its successor, the more abbreviated demotic, and the Greek.

Despite the lack of a cosmic Rosetta Stone, some mathematicians believe that a language suitable for radio communication between different worlds can be evolved through the use of symbolic logic. The effort to develop a purely logical language, in which all mathematical reasoning could be expressed, began at the end of the nineteenth century with the work of several mathematicians, in particular Giuseppe Peano in Italy. In 1900 a lecturer in mathematics at Trinity College, Cambridge, went to Paris for a mathematical congress, bringing with him one of his most brilliant students. The lecturer was named Alfred North Whitehead and his companion was Bertrand Russell. They were greatly excited by news of Peano's new application of logic and ultimately decided to collaborate in carrying it further. The fruit of this collaboration was their monumental three-volume work *Principia Mathematica*, published between 1910 and 1913. Thereafter, Whitehead went to Harvard and became one of the foremost philosophers of his time. His collaborator became the third Earl Russell, philosopher, winner of a Nobel Prize in Literature and, even in his ninetieth year, a gaunt, frail figure demonstrating in Trafalgar Square against nuclear weapons testing.

Their work on 'logistic language' had been preceded by efforts to produce an international language that would bring the world closer together. The first to be widely used was Volapük, invented in 1880 by an Austrian priest. This was followed seven years later by Esperanto, but the mathematician Peano felt these had failed to escape from the arbitrary and illogical syntax of tongues that had evolved in the chance manner of nature. In 1903 he produced his Interlingua, derived from classical Latin but with a simplified syntax. It is still widely used in abstracting scientific articles.

These developments have led Hans Freudenthal, Professor of Mathematics at the University of Utrecht, to attempt extending the 'logistic language' of Whitehead and Russell into something intelligible to beings with whom we have nothing in common except intelligence. He calls it 'Lincos', as a short form of 'Lingua Cosmica'. The logical exposition of the language, as might take place in an extended interstellar message, is contained in his book, *Lincos: Design of a Language for Cosmic Intercourse*, published in the Netherlands in 1960. Actually, he pointed out, such a language already may be established as the vehicle for cosmic intercourse. 'Messages in that language might unceasingly travel through the universe,' he said.

His language is 'articulated', so to speak, by unmodulated radio signals of varying duration and wavelength. These represent 'phonemes' which can be combined to constitute concepts or words. The lexicon and syntax are built up gradually, much in the way that a first-grade reader uses a new word over and over in various ways, then introduces another word: 'See John run. John runs to Mary . . .', etc.

Freudenthal's book-length 'language lesson', like Morrison's simpler approach, begins with elementary arithmetical concepts, but it soon passes on to more abstract ideas. Thus, at an early stage, the idea of right and wrong is conveyed. This is done by presenting a series of mathematical statements, all of which are obviously correct. Each of them is followed by the new word: 'good'. Then a mistaken statement is given, followed by another new word: 'bad'. Later such abstract concepts as honesty, lying, understanding and the like are presented. The teaching process uses 'actors' or 'voices', somewhat in the Socratic method, their dialogues illustrating the meanings of new words.

To represent the radio signals in writing, Freudenthal uses a mixture of mathematical, biological and linguistic symbols including some of those employed earlier by Whitehead and Russell. He points out that Lincos has not yet been developed to where it can explain the diversity of human individuals, but he does use it to express some of the information touched on in Drake's pictorial message; namely, the manner of human reproduction. This is set forth, in Lincos, as follows:

Ha Inq *Ha*▮
x ∈ Hom.→:Ini.x Ext·−:Ini·Cor x.Ext· =
Cca.Sec 11 × 10^{10111}⁝
⋁x⁝x ∈ Bes.∧:Ini.x Ext·−:Ini:Cor x.Ext· > Sec 0⁝
x ∈ Hom.→·⋁⌜y.z⌝:y⩘z ∈ Hom.∧·y = .Mat x·∧:z =
Pat x⁝
⋁x:x ∈ Bes.∧·⋁⌜y.z⌝:y⩘z ∈ Bes.∧·y = .Mat x·∧·z =
Pat x⁝
x ∈ Hom.→⁝⋀t:Ini·Cor x.Ext:Ant:t:Ant:Ini.x Ext·
→:t Cor x.Par·t Cor.Mat x⁝
⋁x:x ∈ Bes.∧·⋀t.Etc⁝
x ∈ Hom.∧:s = Ini·Cor x.Ext·
→:⋁⌜u.v⌝s Cor u.Par·s Cor.Mat x:
∧·Pau Ant.s·Cor v:Par:Pau Ant.s·Cor.Pat x:
∧:s Cor x.Uni·s Cor u.s Cor v⁝
⋁x:x ∈ Bes.∧.Etc⁝
Hom = Hom Fem.∪.Hom Msc⁝
Hom Fem ∩ Hom Msc = ⌜⌝⁝
Car: ↑ x·Nnc x Ext.∧.x ∈ Hom Fem·
Pau > ·Car: ↑ x·Nnc x Ext.∧.x ∈ Hom Msc⁝
y = Mat x.∧.y ∈ Hom·→.y ∈ Hom Fem:∧:
y = Pat x.∧.y ∈ Hom·→·y ∈ Hom Msc⁝
x ∈ Hom ∪ Bes:→:Fin.Cor x·Pst.Fin x ⋕

The three-letter symbols are derived from Latin roots. For example, ‘Fem’ means ‘female’; ‘Msc’ means ‘male’. In his ‘summary’ of this passage, Freudenthal translates it, in part, as follows:

The existence of a human body begins some time earlier than that of the human itself. The same is true for some animals. Mat, mother. Pat, father. Before the individual existence of a human, its body is part of the body of its mother. It has originated from a part of the body of its mother and a part of the body of its father. . . .

Freudenthal comments cryptically that the last paragraph of the coded text ‘is somewhat premature. Its notions,’ he continues, ‘will not be used in this volume.’ He believes that television, including a three-dimensional form, will come only at an advanced stage of interstellar communication, since analytic geometry will be necessary to explain the method. This he considers too advanced for beginners in cosmic discourse. His approach is thus fundamentally different from that of Drake, Morrison and others, who feel instructions for television contact could be conveyed as the essential first step. In fact, the usefulness of Lincos has been seriously questioned by other mathematicians, such as Hogben, who consider it doubtful that what seems purely logical to us would appear so to distant minds, schooled in utterly different methods of thought.

19

What If We Succeed?

ON the night of 30 October 1938 there were manifestations of panic in widely scattered parts of the United States. They began shortly after Orson Wells, the actor and producer, sat in front of a microphone in the New York studios of the Columbia Broadcasting System to introduce an adaptation of *The War of the Worlds*, written in 1898 by H. G. Wells. Orson Welles told his nationwide audience how, in recent years, people on earth had gone about their daily lives in complacence. 'Yet across an immense ethereal gulf, minds that are to our minds as ours are to the beasts in the jungle, intellects vast, cool and unsympathetic regarded this earth with envious eyes and slowly and surely drew their plans against us . . .'

The dramatization that followed, consisting of simulated news bulletins, interviews and sometimes fearful sound effects, was so realistic that thousands believed Martians had, in fact, landed in New Jersey – hideous creatures that slew all opposing them with a sinister 'heat-ray'. People rushed into the streets only partially clothed or struck out aimlessly across open country. Cars raced wildly through crowded cities.

Is this a hint of what might happen if we did, in fact, make contact with a superior civilization? Possibly so, according to a report submitted to the Federal government in 1960. For so dramatic an effect, the encounter would presumably have to be physical, but even radio contact would lead to profound upheavals, the report said. The document was a by-product of the historic action taken by Congress on the heels of the first Sputniks in establishing an agency for the exploration of space. The National Aeronautics and Space Act of 29 July 1958, called for 'long-range studies' of the benefits and problems to be expected from space activities. Pursuant to this act, NASA set up a Committee on Long-Range Studies and awarded a study contract to the Brookings Institution. More than 200 specialists were interviewed by a team led by Donald N. Michael. a social psychologist who later became Director of the Peace Research Institute in Washington. Pertinent portions of the resulting report were reviewed by such figures as Lloyd V. Berkner, head of the Space Science Board, Caryl P. Haskins, President of the Carnegie Institution of Washington, James R. Killian, Chairman of the Corporation of M.I.T., Oscar Schachter, Director of the General Legal Division of the United Nations, and Margaret Mead, the anthropologist.

The document was submitted to NASA only a few months after Project Ozma's attempt to intercept signals from two nearby stars, and much in the minds of those who drafted it was the question of what would happen if we discovered another, far more advanced civilization. The report did not rule out the possibility of direct contact, such as the one so vividly dramatized by Orson Welles, and it suggested that artifacts left by explorers from another world 'might possibly be discovered through our space activities on the Moon, Mars, or Venus.' Nevertheless, it said, if intelligent life is discovered beyond the earth during the next twenty years, it most probably will be in a distant solar system and manifest itself by radio. Such circumstances, it added, would not necessarily rule out revolutionary effects:

> Anthropological files [the report said] contain many examples of societies, sure of their place in the universe, which have disintegrated when they have had to associate with previously unfamiliar societies espousing different ideas and different life ways; others that survived such an experience usually did so by paying the price of changes in values and attitudes and behaviour.
>
> Since intelligent life might be discovered at any time via the radio telescope research presently under way, and since the consequences of such a discovery are presently unpredictable because of our limited knowledge of behaviour under even an approximation of such dramatic circumstances, two research areas can be recommended:
>
> 1. Continuing studies to determine emotional and intellectual understanding and attitudes – and successive alterations of them if any – regarding the possibility and consequences of discovering intelligent extraterrestrial life.
>
> 2. Historical and empirical studies of the behaviour of peoples and

their leaders when confronted with dramatic and unfamiliar events or social pressures . . .

Such studies, the report continued, should consider public reactions to past hoaxes, 'flying saucer' episodes and incidents like the Martian invasion broadcast. They should explore how to release the news of an encounter to the public – or withhold it, if this is deemed advisable. The influence on international relations might be revolutionary, the report concluded, for the discovery of alien beings 'might lead to a greater unity of men on earth, based on the "oneness" of man or on the age-old assumption that any stranger is threatening'.

Much would depend, of course, on the nature of the contact and the content of any message received. Man is so completely accustomed to regarding himself as supreme that to discover he is no more an intellectual match for beings elsewhere than our dogs are for us would be a shattering revelation. Carl Gustav Jung, the disciple of Freud who later went his own psychological way, has said of a direct confrontation with such creatures: The 'reins would be torn from our hands and we would, as a tearful old medicine man once said to me, find ourselves "without dreams", that is, we would find our intellectual and spiritual aspirations so outmoded as to leave us completely paralysed.'

There has been some debate recently in scientific circles as to whether our new acquaintances would, in fact be 'nice people' or the monstrous villains depicted by Orson Welles. Much of the population has been conditioned by science-fiction tales of evil genius at work among the stars, of death rays and battles between galaxies. This may, in part, have accounted for the reaction to the Welles broadcast. Some of those concerned with the search for life in other worlds have sought to counter this attitude.

Thus Philip Morrison, in one of his lectures, questioned whether any civilization with a superior technology would wish to do harm to one that has just entered the community of intelligence. If he were looking through a microscope, he said, and saw a group of bacteria spell out, like a college band, 'Please do not put iodine on this plate. We want to talk to you,' his first inclination, he said, would certainly not be to rush the bacteria into a sterilizer. He doubted that advanced societies 'crush out any competitive form of intelligence, especially when there is clearly no danger.'

Ronald Bracewell, in his lecture at the University of Sydney, asked whether beings in another world would covet our gold or other rare substances. Do they want us as cattle or as slaves? He replied by pointing out the literally astronomical cost of transport between solar systems. Any civilization able to cover interstellar distances would hardly need us for food or raw material, which they could far more easily synthesize at home. 'The most interesting item to be transferred from star to star,' he said, 'is information, and this can be done by radio.'

Edward Purcell, who forcefully argued the impracticality of travel between solar systems, said the relations between such widely separated civilizations must be 'utterly benign'. Maybe journeys over such distances could be made 'by magic', he added, 'but you can't get there by physics.'

Arthur C. Clarke, one of the most knowledgeable of all science-fiction writers, has pointed out the difficulty of administering a galactic empire, be it benign or tyrannical, because the distances would make communications so very slow. Radio messages to our nearest neighbours would take a decade or more.

Nevertheless there has been at least one dissent in the scientific community and its author, as might be expected, is among those who refuse to dismiss the possibility of travel between solar systems, particularly if one does not insist that the trip be completed within a human lifetime. The dissenter was Freeman Dyson, the brilliant young physicist at the Institute for Advanced Study in Princeton. In a letter to *Scientific American* in the spring of 1964 he questioned whether one was justified in assuming that the distant creatures with whom we may converse are moral by our standards.

'Intelligence may indeed be a benign influence,' he said, 'creating isolated groups of philosophy-kings far apart in the heavens and enabling them to share at leisure their accumulated wisdom.' On the other hand, he added, 'intelligence may be a cancer of purposeless technological exploitation, sweeping across a galaxy as irresistibly as it has swept across our own planet.' Assuming interstellar travel at moderate speeds, he said, 'the technological cancer could spread over a whole galaxy in a few million years, a time very short compared with the life of a planet'.

What our detectors will pick up is a technological civilization, he argued, but it will not necessarily be intelligent, in the pure sense of the word. In fact, he continued, it may even be that the society we are inherently likely to detect is more probably 'a technology run wild, insane or cancerously spreading than a technology firmly under control and supporting the rational needs of a superior intelligence'. It is possible that a 'truly intelligent' society might no longer feel the need of, or be interested in, technology.

Our business as scientists is to search the universe and find out what is there. What is there may conform to our moral sense or it may not. ... It is just as unscientific to impute to remote intelligences wisdom and serenity as it is to impute to them irrational and murderous impulses. We must be prepared for either possibility and conduct our searches accordingly.

This point of view is reflected in the questions of those who ask: Should we reply, if we hear someone calling? Since message travel times are long, a hasty reply would be uncalled for, but it seems to this writer hard to believe that we would hear from a race of villains or a civilization run amok. As noted in the chapter entitled 'Is There Intelligent Life on Earth?' an achievement of lasting peace and stability seems to be an important element in qualifying for membership in the interstellar community. While Dyson is understandably dismayed at what technology and population growth are doing to the face of our planet, a society that could not bring such trends under control would probably suffer the disintegration envisioned by Fred Hoyle.

In fact, the achievement of great stability and serenity would seem a prerequisite for a society willing to make the expensive and enormously prolonged effort required to contact another world.

What, indeed, do we know about the roots of evil – of greed, aggressiveness and treachery? On earth they are clearly manifestations of a complex society that has not reached stability. Animals, as a rule, are aggressive only when necessary. The well-fed lion lies near the water-hole while his traditional victims, sensing his satiety, quench their thirst in peace. Man's inherent aggressiveness has made wars possible – though it does not necessarily initiate wars. But it would appear that, if man is to survive, this aspect of his personality will have to be controlled – an achievement that in the past few years has seemed increasingly possible.

Because we have information on only one form of intelligent, technological life we tend to think of such life elsewhere as resembling ourselves more closely than is probably justified. The creatures in which we are interested, besides having minds, must be able to move about and to build things. That is, they must have something comparable to hands and feet. They must have senses, such as sight, touch and hearing, although the senses that evolve on any given planet will be determined by the environment. For example, it may be that, for various reasons, vision in the infra-red part of the spectrum will be more useful than sight in the wavelengths visible to human eyes. Creatures fulfilling such requirements might bear little resemblance to man. As Philip Morrison has put it, they may be 'blue spheres with twelve tentacles'. They may be as big as a mountain or as small as a mouse, although the amount of food availabe would set a limit on largeness and the fixed sizes of molecules must limit the extent to which the size of a complex brain can be compressed. Life spans in some worlds might be extremely great. The cells of our bodies (with a few exceptions, such as brain cells) are constantly replenishing themselves. It would seem that, barring accident or disease, this should continue indefinitely, but because of some subtle influence the replacement process is imperfect. This, the essence of ageing, is now under intensive study. It is not inconceivable that it can be controlled. Progress in the transplantation of human organs and the manufacture of other body components (such as heart valves) is such that even the most sober medical men believe we may ultimately be able to extend lifetimes considerably. Lives measured in many centuries instead of barely one would make the slowness of interstellar signalling far more acceptable. Yet even life spans comparable to our own of today should not divert our interest in message exchanges. 'Imagine,' says Edward

Purcell of Harvard, 'that a reply to one of your messages was scheduled to be received forty years from now. What a legacy for your grandchildren.'

Otto Struve and others have said that the current awakening of mankind to the possible existence of intelligent life in other worlds is as much a challenge to established ways of thought as was the Copernican revolution that displaced the earth from the centre of the universe. The latter set in motion a religious and philosophical upheaval that only in recent decades has run its course. It will be recalled that Giordano Bruno, the Copernican protagonist, was burned at the stake and Martin Luther argued that, since Joshua, in the Bible, commanded the sun – not the earth – to stand still, the system of Copernicus must be false.

By the mid-nineteenth century great changes had taken place. Father Angelo Secchi, the great Jesuit astronomer, was helping to lay the foundations of modern astrophysics – and worrying about the religious implications of the vast universe opening before him. Could it be, he asked, that God populated only one tiny speck in this cosmos with spiritual beings: '. . . It would seem absurd to find nothing but uninhabited desert in these limitless regions,' he wrote. 'No! These worlds are bound to be populated by creatures capable of recognizing, honouring and loving their Creator.'

Both Protestant and Catholic theologians pursued this subject during the early decades of the twentieth century, many of them viewing it in terms of man's spiritual history as set forth in their dogma. It was proposed by Catholic authorities that beings in other worlds could be in a variety of 'states'.

For example, they might be in a state of grace, such as that enjoyed by Adam and Eve before they succumbed to the serpent's temptation. They might be in the fallen state of mankind after the expulsion from Eden. They might have undergone these stages and been redeemed by some action of God. They might be in a 'state of integral nature', midway between man and angel and immune to death; or they might have proved so evil that redemption was denied them. C. S. Lewis, the Anglican lay theologian, has proposed that the vast distances between solar systems may be a form of divine quarantine: 'They prevent the spiritual infection of a fallen species from spreading'; they block it from playing the role of the serpent in the Garden of Eden. Father Daniel C. Raible of Brunnendale Seminary of the Society of the Precious Blood in Canton, Ohio, said in the Catholic weekly *America* that intelligent beings elsewhere 'could be as different from us, physically, as an elephant is from a gnat.' To be 'human' in the theological sense they need only be 'composites of spirit and matter'. As God could create billions of galaxies, he reasoned, 'so He could create billions of human races each unique in itself.' To redeem such races, he said, God could take on any bodily form. 'There is nothing at all repugnant in the idea of the same Divine Person taking on the nature of many human races. Conceivably, we may learn in heaven that there has been not one incarnation of God's son but many.'

The idea of God incarnate in some strange, if not grotesque, creature is to some Catholics unacceptable. Such a point of view was expressed by Joseph A. Breig, a Catholic journalist, in a dialogue published in a 1960 issue of *America*. There can have been only one incarnation, one mother of God, one race 'into which God has poured His image and likeness', he said.

In reply Father L. C. McHugh, an associate editor of the magazine, wrote: 'Does it not seem strange to say that His power, immensity, beauty and eternity are displayed with lavish generosity through unimaginable reaches of space and time, but that the knowledge and love which alone give meaning to all this splendour are confined to this tiny globe where self-conscious life began to flourish a few millennia ago?'

Like the Catholics, the Protestants have been concerned with whether or not God could have taken on bodily form elsewhere. Paul Tillich, one of the foremost Protestant theologians, argues that there is no real reason why such incarnations could not have taken place: 'Incarnation is unique for the special group in which it happens, but it is not unique in the sense that other singular incarnations for other unique worlds are excluded. . . . Man cannot claim to occupy the only possible place for Incarnation. ... The manifestation of saving power in one place implies that saving power is operating in all places.'

Astronomy tells us that new worlds are constantly evolving and old ones are becoming uninhabitable. The lifetime of our own world is thus limited, he said, but this 'leaves open other ways of divine self-manifestation before and after our historical continuum.' Elsewhere he wrote, with characteristic boldness: 'Our ignorance and our prejudice should not inhibit our thoughts from transcending our earth and our history and even our Christianity.'

The hint of Tillich that, though worlds come and go, spiritual life somewhere in the universe goes on for ever has been carried further by John Macquarrie of the University of Glasgow, in lectures that he gave in 1957 at Union Theological Seminary in New York. He took his cue from the 'steady state' theory proposed by Fred Hoyle, Hermann Bondi and Thomas Gold in which, although the universe is constantly expanding, there is continuous creation of new matter to fill the resulting voids. 'It might well be the case,' he said, 'that the universe has produced and will continue to produce countless millions of . . . histories analogous to human history.' The creation would be cyclical, in that it produces 'the same kind of thing over and over again in endless variations.

'To attempt to draw ultimate conclusions about God and the universe from a few episodes of the history which has been enacted on this planet,' he continued, 'would seem to be a most hazardous if not impossible proceeding.'

Long before Project Ozma the British mathematician and physicist Edward A. Milne proposed that only one incarnation – that of Jesus – was necessary, because, when news of it reached other worlds via radio, they too would be saved. This was challenged by E. L. Mascall, University Lecturer in the Philosophy of Religion at Oxford, who said redemption was, in effect, not exportable. Like a number of other theologians who have taken part in this dialogue, he cited the argument of Saint Thomas Aquinas in the thirteenth century that the Incarnation could have taken one of several forms: instead of God the Son appearing on earth, he said, it might have been God the Father or God the Holy Ghost. He reasoned that none of these alternatives was ruled out on theological grounds. If so, today's scholars argue, why could there not also be incarnations elsewhere?

At the conclusion of his analysis, Mascall was almost apologetic for discussing such hypothetical problems: 'Theological principles tend to become torpid for lack of exercise,' he said, 'and there is much to be said for giving them now and then a scamper in a field where the paths are few and the boundaries undefined; they do their day-to-day work all the better for an occasional outing in the country.'

Actually, such 'outings' have been taken by churchmen as sober as William Ralph Inge, Dean of St Paul's Cathedral in London and known as 'the gloomy dean' for his criticism of modern life. Like Father Secchi, he considered it intolerably presumptuous to believe that spiritual life exists only on this one planet. In a series of lectures given at Lincoln's Inn Chapel in 1931–3 he discussed Plato's concept of a 'Universal Soul', and its parallels with Christian views on the Holy Spirit. This, he said,

> raises the question whether there is soul-life in all parts of the universe. I do not think we need follow Fechner in believing that every heavenly body has a soul of its own. [Gustav Theodor Fechner, the German experimental psychologist, preached a highly animistic philosophy.] But I would rather be a star-worshipper than believe with Hegel (if he was not merely speaking impatiently, which is likely enough) that the starry heavens have no more significance than a rash on the sky, or a swarm of flies. There may be, and no doubt are, an immense number of souls in the universe, and some of them may be nearer to the divine mind than we are.

It is perhaps, a sign of the times and of the convergence of Catholicism and Protestantism that theologians of both faiths have been moved by the discussion of the Incarnation in a poem, 'Christ in the Universe', by the English turn-of-the-century poetess Alice Meynell:

> . . . No planet knows that this
> Our wayside planet, carrying land and wave
> Love and life multiplied, and pain and bliss,
> Bears, as chief treasure, one forsaken grave . . .
>
> But, in the eternities,
> Doubtless we shall compare together, hear
> A million alien Gospels, in what guise
> He trod the Pleiades, the Lyre, the Bear.

O, be prepared, my soul!
To read the inconceivable, to scan
The million forms of God those stars unroll
When, in our turn, we show to them a Man.

There are some faiths that have long preached the existence of many worlds, and so for them the shock of discovery would be lessened. Notable among these are the Buddhists and Mormons. The doctrine of the latter, the Church of Jesus Christ of Latter-Day Saints, is set forth in a series of revelations including one entitled the 'Visions of Moses, as revealed to Joseph Smith the Prophet, in June, 1830'. This tells how Moses 'beheld many lands; and each land was called earth, and there were inhabitants on the face thereof'. God told Moses:

'And worlds without number have I created. . . . But only an account of this earth, and the inhabitants thereof, give I unto you. For behold, there are many worlds that have passed away by the word of my power. And there are many that now stand, and innumerable are they unto man; but all things are numbered unto me, for they are mine and I know them. . . . And as one earth shall pass away, and the heavens thereof even so shall another come; and there is no end to my works, neither to my words.'

This is strikingly similar to the 'steady state' concept advanced by such men as Tillich and Macquarrie.

The holy books of Buddhism, in their own glittering way, speak of many worlds. Such a book is the *Saddharma-Pundarika* or *Lotus of the True Law*. The Lord appears before an assembly of Bodhisattvas, or wise men, gathered from countless worlds and numbering eight times the grains of sand in the river Ganges. He tells them of many worlds, of golden people, jewelled trees, perfumed winds, wafting showers of petals. To illustrate the great number of worlds, he says:

Let there be the atoms of earth of fifty hundred thousand myriads of kotis of worlds; let there exist some man who takes one of those atoms of dust and then goes in an eastern direction fifty hundred thousand myriads of kotis of worlds further on, there to deposit that atom of dust; let in this manner the man carry away from all those worlds the whole mass of earth, and in the same manner, and by the same act as supposed, deposit all those atoms in an eastern direction. Now, would you think, young men of good family, that any one should be able to imagine, weigh, count, or determine [the number of] those worlds?

While the Hindus believe in the transmigration of souls into other bodies or spiritual states, their various cosmologies do not envisage other inhabited worlds. As for the Confucians, their attention has been centred on earthly matters, in particular the behaviour of man, even though some early Chinese scholars spoke of other earths with their own heavens shining upon them.

About a dozen centuries ago Jewish thinkers conceived of the cosmos in somewhat of a steady-state manner. This is reflected in one of the commentaries on the Bible, prepared long ago by scholarly Rabbis and entitled the *Midrash Rabba*. It states that 'The Holy One, blessed be he, builds worlds and destroys them.' However, according to contemporary Jewish theologians there has not been much speculation on the religious implications of extraterrestrial life.

While the Brookings report to NASA dwells on the stresses and strains that would follow our making contact with another planet, it is staggering to consider the potential gains. Life in another world would presumably be at a far more advanced stage of evolution – technological, biological and medical – than ours. Knowledge of such a civilization, its discoveries, its techniques might enable us to leap-frog thousands or even millions of years ahead along the path before us. The agonies of premature death, congenital deformity, insanity, as well as such diseases of our society as prejudice, hatred and war, might be eliminated long before otherwise possible. No matter how much another world differed from our own in superficial ways, we might learn from its history enough to understand better our own: what, for example, brought about the great revolutions in the development of life on earth, the extinction of the giant reptiles, the emergence of the mammals and finally of man.

Alastair Cameron, the astrophysicist, in the introduction to his anthology on interstellar communication, describes the possibility of life in other worlds as 'currently the greatest question in scientific philosophy'. Already, he says, we are admitting that there may be millions of societies more advanced than ourselves in our galaxy alone. If we can now take the next step and communicate with some of these societies, then we can expect to obtain an enormous enrichment of all phases of our sciences and arts. Perhaps we shall also receive valuable lessons in the techniques of stable world government.

Approaching the problem from a theological viewpoint, Pierre Teilhard de Chardin, the Jesuit scholar and philosopher, came to a similar conclusion. In *The Phenomenon of Man*, one of the works that, because of their unorthodoxy, remained unpublished until his death in 1955, he discussed the 'mutual fecundation' that would occur if two civilizations that had evolved independently should make contact across space. While the possibility of such an encounter was remote, he said, it would enormously enrich both societies.

One question that has been raised by J. Robert Oppenheimer, the nuclear physicist, is whether or not the science of another civilization would be comprehensible to us. He cites the 'uncertainty principle' first recognized a generation ago, which says that we have a choice as to which traits of an atomic system we wish to study and measure, but in any single case we cannot measure them all. For example, in an experiment, we can determine the location of a particle but, if we do so, we cannot find out its speed. Similarly our science has concentrated on asking certain questions at the expense of others, although this is so woven into the fabric of our knowledge that we are generally unaware of it. In another world the basic questions may have been asked differently.

Thus, while the period of Isaac Newton taught us that scientific truth in the most distant corner of the universe is identical to that in any laboratory on earth, yet in this century it has been shown that there are various ways in which we can seek this truth, and science in one world may be seeking it by means different from those in another. This would make it more difficult to understand the work of our distant neighbours, but it seems unlikely that such understanding would be impossible.

Most exciting of all the prospects are the spiritual and philosophical enrichment to be gained from such exchanges. Our world is undergoing revolutionary changes. Our ancestors enjoyed a serenity denied to most of us. As we devastate our planet with industrialization, with highways, housing and haste, the restoration of the soul that comes from contemplating nature unmarred by human activity becomes more and more inaccessible. Furthermore, through our material achievements, we are threatened by what the French call *embourgeoisement* – domination of the world by bourgeois mediocrity, conformity and comfort-seeking.

The world desperately needs a global adventure to rekindle the flame that burned so intensely during the Renaissance, when new worlds were being discovered on our own planet and in the realms of science. Within a generation or less we will vicariously tread the moon and Mars, but the possibility of ultimately 'seeing' worlds in other solar systems, however remote, is an awesome prospect. 'The soul of man was made to walk the skies,' the English poet Edward Young wrote in the eighteenth century:

> how great,
> How glorious, then, appears the mind of man,
> When in it all the stars and planets roll!
> And what it seems, it is. Great objects make
> Great minds, enlarging as their views enlarge;
> Those still more godlike, as these more divine.

The realization that life is probably universal, however thinly scattered through the universe, has meaning for all who contemplate the cosmos and the mortality of man. George Wald, Professor of Biochemistry at Harvard and a long-time participant in the search for an understanding of life's origins, has said:

> Life has a status in the physical universe. It is part of the order of nature. It has a high place in that order, since it probably represents the most complex state of organization that matter has achieved in our universe. We on this planet have an especially proud place as men; for in us as men matter has begun to contemplate itself . . .

Harlow Shapley, the astronomer, has likewise found spiritual wealth in the new discoveries. They have contributed, he said, 'to the unfolding of a magnificent universe':

> To be a participant is in itself a glory. With our confreres on distant planets; with our fellow animals and plants of land, air, and

sea; with the rocks and waters of all planetary crusts, and the photons and atoms that make up the stars – with all these we are associated in an existence and an evolution that inspires respect and deep reverence. We cannot escape humility. And as groping philosophers and scientists we are thankful for the mysteries that still lie beyond our grasp.

The universality of life also has meaning with regard to the mortality, not only of individuals on earth, but of our planet itself. Bertrand Russell has pointed out that 'all the labours of the ages, all the devotion, all the inspirations, all the noonday brightness of human genius, are destined to extinction in the vast death of the solar system'. Yet it seems that life, in a sense, may be eternal. Perhaps true wisdom is a torch – one that we have not yet received, but that can be handed to us by a civilization late in its life and passed on by our own world as its time of extinction draws near. Thus, as our children and grandchildren offer some continuity to our personal lives, so our communion with cosmic manifestations of life would join us with a far more magnificent form of continuity.

The Spanish-American philosopher George Santayana, in his preface to a volume of Spinoza's works, discussed a form of immortality enjoyed by those who commune with the eternal: 'He who, while he lives, lives in the eternal, does not live longer for that reason. Duration has merely dropped from his view; he is not aware of or anxious about it; and death, without losing its reality, has lost its sting. The sublimation of this interest rescues him, so far as it goes, from the mortality which he accepts and surveys.'

The universe that lies about us, visible only in the privacy, the intimacy of night, is incomprehensibly vast. Yet the conclusion that life exists across this vastness seems inescapable. We cannot yet be sure whether or not it lies within reach, but in any case we are a part of it all; we are not alone!

References

1. Order of the Dolphin

Pearman, J. P. T., 'Extraterrestrial Intelligent Life and Interstellar Communication: An Informal Discussion', Item 28 in A. G. W. Cameron (ed.), *Interstellar Communication, A Collection of Reprints and Original Contributions*, W. A. Benjamin, Inc. (New York, 1963).

United States Congress, House of Representatives, Committee on Science and Astronautics. *Panel on Science and Technology*, 87th Cong., 2nd Sess. Hearings, 4th Mtg., 22 March 1962 (Wash., D.C.), pp. 73–4.

2. Spheres within Spheres

Flammarion, Camille, *La Pluralité des Mondes Habités, Étude où l'on expose les conditions d'habitabilité des terres célestes discutées au point de vue de l'astronomie, de la physiologie et de la philosophie naturelle*, 2nd ed. (Paris, 1864). Book One deals with early speculations.

Heath, Sir Thomas, *Aristarchus of Samos, the Ancient Copernicus*, Clarendon Press (Oxford, 1913).

Laertius, Diogenes, *Lives of Eminent Philosophers*, with an English translation by R. D. Hicks, Harvard Univ. Press (Cambridge, Mass, 1931–50). See sections on Anaxagoras, Zeno and other early thinkers.

Lucretius, *On the Nature of the Universe*, translated with an introduction by Ronald Latham, Penguin Books, (1951), pp. 29, 91, 92.

Munitz, M. K., ed., *Theories of the Universe from Babylonian Myth to Modern Science*, Allen & Unwin (1958). See texts of relevant documents from ancient, Renaissance and post-Renaissance periods.

Plato, *Timaeus*, from *The Collected Dialogues of Plato*, Edith Hamilton and H. Cairns, eds., Bollingen Series 71, Pantheon (New York, 1961), pp. 1164–5.

Robertson, H. P., 'The Universe', *Scientific American*, vol. 195 (Sept. 1956), p. 73.

Teng Mu, quoted by Philip Morrison, *Bulletin of the Philosophical Soc. of Wash.*, vol. 16 (1962), p. 81.

3. Science Reborn

Borel, Pierre, quoted by Flammarion, op. cit., pp. 30–31.

Bruno, Giordano: see *Giordano Bruno, His Life and Thought, With Annotated Translations of His Work, On the Infinite Universe and Worlds*, by Dorothea Waley Singer, Henry Schuman (New York, 1950), pp. 257, 304; also *Historical Trials* by Sir John Macdonnell (Oxford, 1927), p. 83; and *Giordano Bruno* by J. L. McIntyre (London, 1903), p. 99.

Cyrano de Bergerac, *Voyage à la lune*, extract in Flammarion, op. cit., pp. 509–14.

Digges, Thomas, quoted by Munitz, op. cit., p. 188.

Dingle, Herbert, 'Cosmology and Science', *Scientific American*, vol. 195 (Sept. 1956), p. 228.

de Fontenelle, Bernard Le Bovier, *Conversations on the Plurality of Worlds* (London, 1809), with a 'Memoir of the Life and Writings of the late Monsieur de Fontenelle', by M. de Voltaire, pp. 5, 59, 78, 98. (Originally published in 1686 as *Entretiens sur la pluralité des mondes*.)

Huygens, Christiaan, *Cosmotheros*, quoted by Munitz, op. cit., pp. 221–2.

Milton, John, *Paradise Lost*, Book Eighth, lines 145 ff.

Newton, Isaac, letter to Richard Bentley, 10 Dec. 1692, quoted by Munitz, op. cit., p. 211.

Pope, Alexander, *An Essay on Man*, Epistle I, lines 19–28.

Wilkins, Bishop John, *Mathematical Magick* (1648), quoted by Vilhjalmur Stefansson in *Northwest to Fortune*, Duell, Sloan & Pearce (New York, 1958), p. 333.

4. Is Our Universe Unique?

Herschel, William, 'On the Construction of the Heavens', see excerpts in Munitz, op. cit., pp. 264–8.

Kant, Immanuel, *Universal Natural History and Theory of the Heavens*, see excerpts in ibid., pp. 231–49

Leavitt, H. S., 'Discovery of the Period-Magnitude Relation', in *Source Book in Astronomy 1900–1950*, H. Shapley, ed., Oxford University Press (1960), pp. 186–9, excerpted from 'Periods of 25 Variable Stars in the Small Magellanic Cloud', *Harvard Circular*, No. 173 (1912).

'1777 Variables in the Magellanic Clouds', *Annals of Harvard College Observatory*, vol. 60, no. 4 (1908), pp. 87–108.

Shapley, Harlow, 'From Heliocentric to Galactocentric', in *Source Book in Astronomy*, op. cit., pp. 319–24, excerpted from *Star Clusters*, Harvard Observatory Monographs, No. 2, McGraw-Hill (New York, 1930), pp. 171–8.

Of Stars and Men, The Human Response to an Expanding Universe, Elek Books (1958), pp. 9, 53, 55.

Struve, Otto, and Velta Zebergs, *Astronomy of the 20th Century*, Collier-Macmillan (1963), Chapters 15 ('Pulsating Variable Stars'), 19 ('The Milky Way') and 20 ('Galaxies').

de Vaucouleurs, Gérard, 'The Supergalaxy', *Scientific American*, vol. 191 (July 1954), pp. 30–35.

Wright, Thomas, *An Original Theory or New Hypothesis of the Universe* (1750), quoted by Munitz, op. cit., p. 226, and by Munitz in *Space, Time and Creation, Philosophical Aspects of Scientific Cosmology*, Collier-Macmillan (1962), p. 22.

5. The Solar System: Exception or Rule?

Cameron, A. G. W., 'The History of Our Galaxy' and 'The Origin of the Solar System', in Cameron, op. cit. (see p. 316), Items 2 and 3.

'The Formation of the Sun and Planets', *Icarus*, vol. 1 (May, 1962), pp. 13–69.

'Studies on the Origin of the Solar System', *Yale Scientific Magazine*, vol. 38 (April, 1963), pp. 12–14.

Eucken, Arnold, 'Über den Zustand des Erdinnern', *Die Naturwissenschaften* (Berlin, April-June, 1944), pp. 112–21.

Flammarion, op. cit., pp. 48–9, 266–9 (on the views of Bode and Kant).

Fowler, W. A., 'Nuclear Clues to the Early History of the Solar System', *Science*, vol. 135 (23 March 1962), pp. 1037–45.

Schmidt, O. J.: see B. J. Levin, 'The Origin of the Solar System', *New Scientist*, vol. 13 (8 Feb. 1962), p. 323.

Struve and Zebergs, op. cit., Chapter 9 ('Origin of the Solar System').

Swedenborg, Emanuel, 'Earths in Our Solar System Which Are Called Planets and Earths in the Starry Heaven Their Inhabitants, and the Spirits and Angels There from Things Heard and Seen', from the Latin of Emanuel Swedenborg, Swedenborg Society (London, 1962).

Ter Haar, D., and A. G. W. Cameron, 'Historical Review of Theories of the Origin of the Solar System', in *Origin of the Solar System, Proceedings of a conference held at the Goddard Institute for Space Studies, New York, 23–24 January 1962*, Robert Jastrow and A. G. W. Cameron, eds., Academic Press (New York, 1963). See other chapters of the book for current thinking on origin of the solar system.

Urey, H. C., *The Planets, Their Origin and Development*, Oxford Univ. Press (1952).

Urey, H. C., 'Diamonds, Meteorites, and The Origin of the Solar System', *Astrophysical Jour.*, vol. 124 (Nov. 1956), pp. 623–37.

Whipple, F. L., 'History of the Solar System', paper presented at the Centennial Meeting, Natnl Acad. of Sciences (Wash., D.C., 21 Oct. 1963).

6. The Puzzle of the 'Slow' Stars

Brown, Harrison, 'Planetary Systems Associated with Main-Sequence Stars', *Science*, vol. 145 (11 Sept. 1964), pp. 1177–81. His argument was criticized in a letter from Andrew T. Young of the Harvard College Observatory in *Science*, vol. 148 (23 April 1965), p. 532.

van de Kamp, Peter, 'Planetary Companions of Stars', *Vistas in Astronomy*, A. Beer, ed., vol. 2, Pergamon Press (London, 1956), pp. 1040–48.

'Barnard's Star as an Astrometric Binary', *Sky and Telescope*, vol. 26 (July 1963), pp. 8–9.

'Astrometric Study of Barnard's Star from Plates Taken with the 24-inch Sproul Refractor', *Astronomical Jour.*, vol. 68 (Sept. 1963), pp. 515–21.

'The Discovery of Planetary Companions of Stars', *Yale Scientific Magazine*, vol. 38 (Dec. 1963), pp. 6–8.

Russell, H. N., 'The Spectrum-Luminosity Diagram', *Source Book in Astronomy*, op. cit., pp. 253–62, excerpted from *Popular Astronomy*, vol. 22 (1914), pp. 275–94, 331–51.

Shajn, G. A., and Otto Struve, 'On the Rotation of the Stars', *Source Book in Astronomy*, op. cit., pp. 116–23, originally in *Monthly Notices of the Royal Astronomical Soc.*, vol. 89 (1929), pp. 222–39.

Spitzer, Lyman, Jr, 'The Beginnings and Future of Space Astronomy', *American Scientist*, vol. 50 (Sept. 1962), pp. 473–84.

Struve, Otto, *Stellar Evolution*, Oxford Univ. Press (1950), p. 130.

'Proposal for a Project of High-Precision Stellar Radial Velocity Work', *The Observatory*, vol. 72 (Oct. 1952), pp. 199–200.

The Astronomical Universe, Condon Lectures, Oregon State System of Higher Education (Eugene, 1958), p. 26.

and Zebergs, op. cit., p. 112, Chapters 10 ('Spectral Classification'), 11 ('Stellar Atmospheres and Spectroscopy'), and 14 ('Double Stars').

7. Where to Look

Brown, Harrison, in *The Atmospheres of the Earth and Planets*, G. P. Kuiper, ed., Univ. of Chicago Press (Chicago, 1952), Chapter 9.

Dole, S. H., and I. Asimov, *Planets for Man*, Blaisdell (Book Centre, 1964).

Huang, S. S., 'A Nuclear-Accretion Theory of Star Formation', *Publications of the Astronomical Soc. of the Pacific*, vol. 69 (Oct. 1957), pp. 427–30.

'The Problem of Life in the Universe and the Mode of Star Formation', ibid., vol. 71 (Oct. 1959), pp. 421–4, reprinted in Cameron, op. cit., as Item 7.

'Life-Supporting Regions in the Vicinity of Binary Systems', ibid., vol. 72 (April 1960), pp. 106–14, reprinted in Cameron, op. cit., as Item 8.

'The Limiting Sizes of the Habitable Planets', ibid., vol. 72 (Dec. 1960), pp. 489–93, reprinted in Cameron, op. cit., as Item 9.

'Occurrence of Life in the Universe', *American Scientist*, vol. 47 (Sept. 1959), pp. 397–402, reprinted in Cameron, op. cit., as Item 6.

'Some Astronomical Aspects of Life in the Universe', *Sky and Telescope*, vol. 21 (June 1961), pp. 312–16.

'Life Outside the Solar System', *Scientific American*, vol. 202 (April 1960), pp. 55–63.

and R. H. Wilson, Jr, 'Astronomical Aspects of the Emergence of Intelligence', paper presented at the Jan. 1963 meeting of the Institute of the Aerospace Sciences (New York), IAS Paper No. 63–48.

Shapley, Harlow, 'Crusted Stars and Self-Heating Planets', *Matemática y Fisica Teórica, Serie A*, vol. 14 (1962), Tucumán (Argentina) National University.

8. Creation or Evolution?

Akabori, Shiro, 'On the Origin of the Fore-protein', *Proceedings of the First International Symposium on the Origin of Life on the Earth*, August 1957, Pergamon Press (London, 1959), pp. 189–96.

Barghoorn, E. S., see 'Woodring Conference on Major Biologic Innovations and the Geologic Record', by P. E. Cloud, Jr, and P. H. Abelson, *Proceedings of the Natnl Acad. of Sciences*, vol. 47 (15 Nov. 1961), pp. 1705–12.

Bernal, J. D., 'The Problem of Stages in Biopoesis', *Proceedings of the First International Symposium* . . . op. cit., pp. 38–53.

Darwin, Charles, letter to G. C. Wallich, 28 March 1882, and an earlier letter from *Notes and Records of the Royal Society of London*, vol. 14, no. 1 (1959), quoted by M. Calvin in *Chemical Evolution*, Condon Lectures, Oregon State System of Higher Education (Eugene, 1961), p. 2.

Dubos, R. J., *Louis Pasteur, Free Lance of Science*, Gollancz (1951), Chapter 6.

Haldane, J. B. S., 'The Origin of Life' (1928), published in *The Rationalist Annual* (1929), pp. 3–10, and in *The Inequality of Man and Other Essays* (London, 1932), pp. 148–60.

'Radioactivity and the Origin of Life in Milne's Cosmology', *Nature*, vol. 153 (6 May 1944), p. 555.

'Origin of Life', *New Biology*, no. 16, Penguin (1954).

van Helmont, J.-B., quoted by Louis Pasteur, *Oeuvres* (Paris, 1922), vol. 2, p. 329.

Michelet, Jules, *La Mer* (Paris, 1861), pp. 116–17.

Oparin, A. I., *Origin of Life*, translated with annotations by Sergius Morgulis, Oliver & Boyd (Edinburgh, 1957). See particularly historical sections in Chapters 1 and 2, and p. 246.

Pasteur, Louis, *Oeuvres*, op. cit., vol. 2, pp. 202–5, 328–46.

Comptes Rendus hebdomadaires des Séances de l'Académie des Sciences (Paris), vol. 50 (1860), pp. 303–7, 849–54; vol. 51 (1860), pp. 348–52; vol. 56 (1863), pp. 734–40.

Pouchet, F. A., ibid., vol. 47 (1858), pp. 979–84; vol. 48 (1859), pp. 148–58, 546–51; vol. 57 (1863), pp. 765–6.

Vallery-Radot, René, *The Life of Pasteur* (London, 1911), vol. 1, pp. 92–112.

9. Building Molecules

Abelson. P. H., 'Amino Acids Formed in "Primitive Atmospheres"' (abstract), *Science*, vol. 124 (9 Nov. 1956), p. 935.

Blum, H. F., 'Perspectives in Evolution', *American Scientist*, vol. 43 (Oct. 1955), pp. 595–610.

'On the Origin and Evolution of Living Machines', ibid., vol. 49 (Dec. 1961), pp. 474–501.

Calvin, Melvin, see 'Reduction of Carbon Dioxide in Aqueous Solutions by Ionizing Radiation' by W. M. Garrison, D. C. Morrison, J. G. Hamilton, A. A. Benson and M. Calvin, *Science*, vol. 114 (19 Oct 1951), pp. 416–18.

'Chemical Evolution and the Origin of Life', *American Scientist*, vol. 44 (July 1956), pp. 248–63.

Calvin, Melvin, 'Origin of Life on Earth and Elsewhere', *Univ. of California Radiation Laboratory Reports*, UCRL 9005 & 9440 (1959, 1960).
Chemical Evolution, Condon Lectures, op. cit.
'Communication: From Molecules to Mars', *AIBS Bulletin*, American Institute of Biological Sciences, vol. 12 (Oct. 1962), pp. 29–44.
and G. J. Calvin, 'Atom to Adam', *American Scientist*, vol. 52 (June 1964), pp. 163–86.
Dubos, op. cit., p. 16.
Fox, S. W., 'A Chemical Theory of Spontaneous Generation', *Proceedings of the First Intnl Symposium . . .*, op. cit., pp. 256–62.
and Kaoru Harada, 'The Thermal Copolymerization of Amino Acids Common to Protein', *Jour. of the Amer. Chemical Soc.*, vol. 82 (20 July 1960), pp. 3745–51.
'How Did Life Begin?' *Science*, vol. 132 (22 July 1960), pp. 200–208.
and Harada, 'Synthesis of Uracil under Conditions of a Thermal Model of Prebiological Chemistry', *Science*, vol. 133 (16 June 1961), pp. 1923–4.
and Shuhei Yuyama, 'Abiotic Production of Primitive Protein and Formed Microparticles', *Annals of the N.Y. Acad. of Sciences*, vol. 108, art. 2 (29 June 1963), pp. 487–94.
Miller, S. L., 'A Production of Amino Acids under Possible Primitive Earth Conditions', *Science*, vol. 117 (15 May 1953), pp. 528–9.
'Production of Some Organic Compounds under Possible Primitive Earth Conditions', *Jour. of the Amer. Chemical Soc.*, vol. 77 (12 May 1955), pp. 2351–61.
and H. C. Urey, 'Organic Compound Synthesis on the Primitive Earth', *Science*, vol. 130 (31 July 1959), pp. 245–51.
'Formation of Organic Compounds on the Primitive Earth', *Proceedings of the First Intnl Symposium . . .*, op. cit., pp. 123–35.
Oró, Joan, A. P. Kimball, R. Fritz and F. Master, 'Amino Acid Synthesis from Formaldehyde and Hydroxylamine', *Archives of Biochemistry & Biophysics*, vol. 85 (Nov. 1959), pp. 115–30.
and A. P. Kimball, 'Synthesis of Purines under Possible Primitive Earth Conditions – I. Adenine from Hydrogen Cyanide', ibid., vol. 94 (Aug. 1961), p. 217.
and C. L. Guidry, 'A Novel Synthesis of Polypeptides', *Nature*, vol. 186 (9 April 1960), pp. 156–7.
'Comets and the Formation of Biochemical Compounds on the Primitive Earth', ibid., vol. 190 (29 April 1961), p. 389.

Oró, Joan and S. S. Kamat, 'Amino-acid Synthesis from Hydrogen Cyanide under Possible Primitive Earth Conditions', ibid., pp. 442–3.

Pavlovskaya, T. E., and A. G. Pasynskii, 'The Original Formation of Amino Acids under the Action of Ultraviolet Rays and Electric Discharges', *Proceedings of the First Intnl Symposium . . .*, op. cit., pp. 151–7.

Ponnamperuma, Cyril, R. M. Lemmon, R. Mariner and M. Calvin, 'Formation of Adenine by Electron Irradiation of Methane, Ammonia, and Water', *Proceedings of the Natnl Acad. of Sciences*, vol. 49 (May 1963), pp. 737–40.

R. Mariner and C. Sagan, 'Formation of Adenosine by Ultraviolet Irradiation of a Solution of Adenine and Ribose', *Nature*, vol. 198 (22 June 1963), pp. 1199–1200.

C. Sagan and R. Mariner, 'Ultraviolet Synthesis of Adenosine Triphosphate Under Possible Primitive Earth Conditions', *Research in Space Science, Special Report No. 128*, Smithsonian Institution Astrophysical Observatory (Cambridge, Mass., 10 July 1963).

Schramm, G., H. Grötsch and W. Pollmann, 'Non-Enzymatic Synthesis of Polysaccharides, Nucleosides and Nucleic Acids and the Origin of Self-Reproducing Systems', *Angewantandte Chemie*, Intnl Ed. (English), vol. 1 (Jan. 1962), pp. 1 7.

Simpson, G. G., 'The Nonprevalence of Humanoids', *Science*, vol. 143 (21 Feb. 1964), pp. 769–75.

Urey, H. C., 'Primitive Planetary Atmospheres and the Origin of Life', *Proceedings of the First Intnl Symposium . . .*, op. cit., pp. 16–22.

Wald, George, 'The Origins of Life', *Proceedings of the Natnl Acad. of Sciences* (Aug. 1964).

Young, R. S., and C. Ponnamperuma, 'Early Evolution of Life', BSCS *Pamphlets*, no. 11 (Boston, 1964), Biological Sciences Curriculum Study, Amer. Inst. of Biological Sciences.

10. Visitors from Space

Anders, Edward, and G. G. Goles, 'Theories on the Origin of Meteorites', *Jour. of Chemical Education*, vol. 38 (Feb. 1961), pp. 58–66.

'Meteorite Ages', *Reviews of Modern Physics*, vol. 34 (April 1962), pp. 287–325.

'Age and Origin of Meteorites', paper presented at annual meeting, Natnl Acad. of Sciences (Wash., D.C., 29 April 1964).

'Origin, Age, and Composition of Meteorites', *Space Science Reviews* (in press).

Arnold, James R., 'Where do Meteorites Come From?' paper presented at the annual meeting, Natnl Acad. of Sciences (Wash., D.C., 29 April 1964).
and S. A. Tyler, 'Fossil Organisms from Precambrian Sediments', *Annals of the N.Y. Acad. of Sciences*, vol. 108, art. 2 op cit., pp. 451–2.
Berzelius, J. J., *Kongl. Vetenskapsacademiens Handlingar* (Stockholm, 1834), p. 148.
Chao, E. C. T., E. M. Shoemaker, B. M. Madsen, 'First Natural Occurrences of Coesite', *Science*, vol. 132 (22 July 1960), pp. 220–22.
Cloëz, S., Daubrée, etc., *Comptes Rendus*, op. cit., vol. 58 (1864), pp. 984–90; vol. 59 (1864), pp. 37–40.
DeFelice, J., G. G. Fazio and E. L. Fireman, 'Cosmic-Ray Exposure Age of the Farmington Meteorite from Radioactive Isotopes', *Science*, vol. 142 (8 Nov. 1963), pp. 673–4.
Frondel, Clifford, 'Mineralogical Problems presented by Meteorites', paper presented at the annual meeting, Natnl Acad. of Sciences (Wash., D.C., 29 April 1964).
Gilbert, L. W., ed. (description of the Alais fall), *Annalen der Physik*, vol. 24 (1806), pp. 189–94.
Gilvarry, J. J., 'Origin and Nature of Lunar Surface Features', *Nature*, vol. 188 (10 Dec. 1960), pp. 886–91.
Goles, G. G., R. A. Fish and E. Anders, 'The Record in the Meteorites—: The Former Environment of Stone Meteorites as Deduced from K^{40}—Ar^{40} Ages', *Geochimica et Cosmochimica Acta*, vol. 19 (July 1960), pp. 177–95.
Jacchia, L. G., 'Meteors, Meteorites, and Comets: Interrelations', Chapter 22 from *The Moon, Meteorites and Comets*, vol. 4 of *The Solar System*, Univ. of Chicago Press (1963).
Kulik, L. A., 'The Tunguska Meteorite', *Source Book in Astronomy 1900–1950*, op. cit., pp. 75–9.
Mason, Brian, 'Origin of Chondrules and Chondritic Meteorites' *Nature*, vol. 186 (16 April 1960), pp. 230–31.
Meteorites, John Wiley & Sons (1962).
Mueller, G., 'The Properties and Theory of Genesis of the Carbonaceous Complex within the Cold Bokkevelt Meteorite', *Geochimica et Cosmochimica Acta*, vol. 4 (Aug. 1953), pp. 1–10.
Urey, H. C., *The Planets*, op. cit., pp. 193–210.
and H. Craig, 'The Composition of the Stone Meteorites and the Origin of the Meteorites', *Geochimica et Cosmochimica Acta*, vol. 4 (Aug. 1953), pp. 36–82.
'Diamonds, Meteorites, and the Origin of the Solar System', op. cit.

Urey, H. C., 'Primary and Secondary Objects', *Jour. of Geophysical Research*, vol. 64 (Nov 1959)., pp. 1721–37.

'The Chemical History of Meteorites', paper presented at the annual meeting, Natnl Acad. of Sciences (Wash., D.C., 29 April 1964).

Wöhler, F., 'Neuere Untersuchungen über die Bestandtheile des Meteorsteines vom Capland', *Sitzungsberichte, Mathematisch-Naturwissenschaftlichen Classe der Kaiserlichen Akademie der Wissenschaften* (Vienna, 1860), vol. 41, pp. 565–7.

Wood, J. A., 'Chondrules and the Origin of the Terrestrial Planets', *Nature*, vol. 194 (14 April 1962), pp. 127–30.

11. 'Wax and Wigglers'

Anders, Edward, 'The Moon as a Collector of Biological Material', *Science*, vol. 133 (14 April 1961), pp. 1115–16.

'Meteoritic Hydrocarbons and Extraterrestrial Life', *Annals of the N.Y. Acad. of Sciences*, vol. 93 (29 Aug. 1962), pp. 651–4.

and F. W. Fitch, 'Search for Organized Elements in Carbonaceous Chondrites', *Science*, vol. 138 (28 Dec. 1962), pp. 1392–9.

'On the Origin of Carbonaceous Chondrites', *Annals of the N.Y. Acad.*, vol. 108, art. 2, op. cit., pp. 514–33; see also 'Panel Discussion', pp. 611–12.

Eugene R. DuFresne, Ryoichi Hayatsu, Albert Cavaillé, Ann DuFresne and Frank W. Fitch, 'Contaminated Meteorite', *Science*, vol. 146 (27 Nov. 1964), pp. 1157–61.

Bergamini, David, 'Wax and Wigglers: Life in Space?' *Life* (5 May 1961), pp. 57–62.

Bernal, J. D., 'The Problem of the Carbonaceous Meteorites', *The Times Science Review* (Summer, 1961), pp. 3–4.

'Comments', in symposium on 'Life-Forms in Meteorites', *Nature*, vol. 193 (24 March 1962), pp. 1127–9.

'Is There Life Elsewhere in the Universe?' *The Listener* (26 April 1962), pp. 723–4.

see 'Panel Discussion', *Annals of the N.Y. Acad.*, vol. 108, art. 2, op. cit., pp. 606–15.

Berthelot, M., *Comptes Rendus*, op. cit., vol. 67 (1868), p. 849.

Bourrelly, Pierre, 'Panel Discussion', op. cit., p. 614.

Briggs, Michael H., 'Organic Constituents of Meteorites', *Nature*, vol. 191 (16 Sept. 1961), pp. 1137–40.

Calvin, M., and S. K. Vaughn, 'Extraterrestrial Life: Some Organic Constituents of Meteorites and Their Significance for Possible Extraterrestrial Biological Evolution', *Space Research: Proceedings of the First International Space Science Symposium*, Hilde Kallmann, ed., North-Holland Publ. Co. (Amsterdam, 1960), pp. 1171–91.

Claus, George, and B. Nagy, 'A Microbiological Examination of Some Carbonaceous Chondrites', *Nature*, vol. 192 (18 Nov. 1961), pp. 594–6.

B. Nagy and D. L. Europa, 'Further Observations on the Properties of the "Organized Elements" in Carbonaceous Chondrites', *Annals of the N.Y. Acad.*, vol. 108, art. 2, op. cit., pp. 580–605.

Farrell, M. A., 'Living Bacteria in Ancient Rocks and Meteorites', *American Museum Novitates*, no. 645, The American Museum of Natural History (New York, 18 July 1933).

Fitch, F. W., and E. Anders, 'Organized Element: Possible Identification in Orgueil Meteorite', *Science*, vol. 140 (7 June 1963), pp. 1097–1100.

and E. Anders, 'Observations on the Nature of the "Organized Elements" and Carbonaceous Chondrites', *Annals of the N.Y. Acad.*, vol. 108, art. 2, op. cit., pp. 495–513.

H. P. Schwarcz and E. Anders, '"Organized Elements" in Carbonaceous Chondrites', *Nature*, vol. 193 (24 March 1962), pp. 1123–5.

Hahn, Otto, *Die Meteorite (Chondrite) und ihre Organismen* (Tübingen, 1880).

Haldane, J. B. S., 'Origin of Life', op. cit., pp. 18–25.

Hodgson, G. W., and B. L. Baker, 'Evidence for Porphyrins in the Orgueil Meteorite', *Nature*, vol. 202 (11 April 1964), pp. 125–31.

Lipman, C. B., 'Are There Living Bacteria in Stony Meteorites?' *American Museum Novitates*, op. cit., no. 588 (31 Dec. 1932).

Meinschein, W. G., 'Hydrocarbons in the Orgueil Meteorite', *Proceedings of Lunar Planetary Exploration Colloquium*, North American Aviation, Inc. (Downey, Calif., 15 Nov. 1961), vol. 2, no. 4, pp. 65–7.

Mueller, G., 'Interpretation of Micro-structures in Carbonaceous Meteorites', *Nature*, vol. 196 (8 Dec. 1962), pp. 929–32.

Nagy, Bartholomew, W. G. Meinschein and D. J. Hennessy, 'Mass Spectroscopic Analysis of the Orgueil Meteorite: Evidence for Biogenic Hydrocarbons', *Annals of the N.Y. Acad.*, vol. 93 (5 June 1961), pp. 25–35.

Meinschein and Hennessy, Discussion of the paper by Anders (cited above), *Annals of the N.Y. Acad.*, vol. 93 (29 Aug. 1962) pp. 658–64.

K. Fredriksson, H. C. Urey, G. Claus, C. A. Andersen and J. Percy, 'Electron Probe Microanalysis of Organized Elements in the Orgueil Meteorite', *Nature*, vol. 198 (13 April 1963), pp. 121–5.

Nagy, Bartholomew and Sister Mary C. Bitz, 'Long-Chain Fatty Acids from the Orgueil Meteorite', *Archives of Biochemistry and Biophysics*, vol. 101 (May 1963), pp. 240–8.

Meinschein and Hennessy, 'Aqueous, Low Temperature Environment of the Orgueil Meteorite Parent Body', *Annals of the N.Y. Acad.*, vol. 108, art. 2, op. cit., pp. 534–52.

K. Fredriksson, J. Kudynowski and L. Carlson, 'Ultra-violet Spectra of Organized Elements', *Nature*, vol. 200 (9 Nov. 1963), pp. 565–6.

M. T. J. Murphy, V. E. Modzeleski, G. Rouser, G. Claus, D. J. Hennessy, U. Colombo and F. Gazzarrini, 'Optical Activity in Saponified Organic Matter Isolated from the Interior of the Orgueil Meteorite', *Nature*, vol. 202 (18 April 1964), pp. 228–33.

Palik, P., 'Further Life-forms in the Orgueil Meteorite', *Nature*, vol. 194 (16 June 1962), p. 1065.

Papp, A., 'Fossil Protobionta and Their Occurrence', *Annals of the N.Y. Acad.*, vol. 108, art. 2, op. cit., p. 462. See also 'Panel Discussion', ibid., p. 613.

Ross, Robert, see 'Panel Discussion', ibid, pp. 608–9.

Sagan, Carl, 'Indigenous Organic Matter on the Moon', and 'Biological Contamination of the Moon', *Proceedings of the Natnl Acad. of Sciences*, vol. 46 (15 April 1960), pp. 393–402.

Sisler, F. D., *Proceedings of Lunar and Planetary Exploration Colloquium*, op cit., pp. 67–73.

Staplin, F. L., 'Microfossils from the Orgueil Meteorite', *Micropaleontology*, vol. 8 (1 July 1962), pp. 343–7.

Sylvester-Bradley, P. C., and R. J. King, 'Evidence for Abiogenic Hydrocarbons', *Nature*, vol. 198 (25 May 1963), pp. 728–31.

Tasch, Paul, see 'Panel Discussion', *Annals of the N.Y. Acad.*, vol. 108, art. 2, op. cit., pp. 612–13.

Timofeyev, B. V., 'On the Occurrence of Organic Remains in Chondritic Meteorites', abstracts of papers presented at the 4th Astrogeological Meeting, Geological Soc. of the USSR, May 1962. See also *Pravda* (Moscow), 11 May 1962, p. 6.

see *Ogonyok*, no. 42 (Moscow, Oct. 1962), pp. 20–21.

Urey, H. C., 'Origin of Life-like Forms in Carbonaceous Chondrites', *Nature*, vol. 193 (24 March 1962), pp. 1119–23.

'Lifelike Forms in Meteorites', *Science*, vol. 137 (24 Aug. 1962), pp. 623–7.

'Panel Discussion', *Annals of the N.Y. Acad.*, vol. 108, art. 2, op. cit., pp. 606–15.

Went, F. W., 'Organic Matter in the Atmosphere, and its Possible

Relation to Petroleum Formation', *Proceedings of the Natnl Acad. of Sciences*, vol. 46 (15 Feb. 1960), pp. 212–21.

12. Is There Life on Mars?

Anders, Edward, and James R. Arnold, 'Age of Craters on Mars', *Science*, vol. 149 (24 Sept. 1965), pp. 1494–6. This is followed by two other reports on this subject (pp. 1496–9).

Anderson, Hugh R., et. al., 'Mariner IV Measurements near Mars: Initial Results', *Science*, vol. 149 (10 Sept. 1965), pp. 1226–48.

Flammarion, Camille, *La Planète Mars et ses Conditions d'Habitabilité* (Paris, 1892).

Gold, Thomas, proposal on radar scanning for signs of life: see F. D. Drake, 'Radio Emission from the Planets', *Physics Today*, vol. 14 (April 1961), p. 34.

Horowitz, N. H., 'The Design of Martian Biological Experiments', *Life Sciences and Space Research*, vol. 2 of *Proceedings of the Symposia on Extraterrestrial Biology and Organic Chemistry, Methods for the Detection of Extraterrestrial Life and Terrestrial Life in Space* – Warsaw, 1963. M. Florkin, ed., Pergamon Press (1966).

Huang, S. S., 'Some Astronomical Aspects of Life in the Universe', op. cit.

Kaplan, L. D., G. Münch and H. Spinrad, 'An Analysis of the Spectrum of Mars', *Astrophysical Jour.*, vol. 139 (1 Jan. 1964), pp. 1–15.

Kuiper, G. P., paper given at annual meeting, Amer. Geophysical Union (Wash., D.C., 1964).

Kuprevich, V. F., Pres., Acad. of Sciences of the Byelorussian SSR, on enriching earth with Martian flora and fauna: 'Earth, Life, Space', *Tekhnika-Molodezhi*, no 9 (Moscow, 1961), p. 11.

Lederberg, J. and C. Sagan, 'Microenvironments for Life on Mars', *Proceedings of the Natnl Acad. of Sciences*, vol. 48 (Sept. 1962), pp. 1473–5.

Levin, G. V., A. H. Heim, J. R. Clendenning, M.-F. Thompson, '"Gulliver" – A Quest for Life on Mars', *Science*, vol. 138 (12 Oct. 1962), pp. 114–21.

Lowell, Percival, *Mars*, Houghton Mifflin (Boston, 1895), pp. 201 and 208.

Miller, S. L., and H. C. Urey, 'Organic Compound Synthesis on the Primitive Earth', *Science*, vol. 130 (31 July 1959), pp. 245–51.

National Academy of Sciences, *A Review of Space Research. The Report of the Summer Study Conducted under the Auspices of the Space Science Board of the National Academy of Science at the State Univ.*

of Iowa, Iowa City, Iowa, 17 June–10 Aug. 1962, NAS Pub. 1079, Chapter 9, pp. 1, 3, 4, 7, 10.

Rea, D. G., T. Belsky and M. Calvin, 'Interpretation of the 3- to 4-Micron Infrared Spectrum of Mars', *Science*, vol. 141 (6 Sept. 1963), pp. 923–7.

Rea, D. G., B. T. O'Leary and W. M. Sinton, 'Mars: The Origin of the 3.58- and 3.69-Micron Minima in the Infrared Spectra', *Science*, vol. 147 (12 March 1965), pp. 1286–8.

Sagan, Carl, 'On the Origin and Planetary Distribution of Life', *Radiation Research*, vol. 15 (Aug. 1961), pp. 174–92.

'Life Beyond the Earth', lecture, 21 May 1962, Voice of America Forum Series on Space Science, *Twenty Lectures*, VOA (Wash., D.C., 1962, mimeo.).

'Biological Exploration of Mars', *Advances in the Astronautical Sciences*, vol. 15, The Amer. Astronautical Soc. (1963), pp. 571–81.

Saheki, Tsuneo, 'Martian Phenomena Suggesting Volcanic Activity', *Sky and Telescope*, vol. 14 (Feb. 1955), pp. 144–6.

Salisbury, F. B., 'Martian Biology', *Science*, vol. 136 (6 April 1962), pp. 17–26.

Shirk, J. S., W. A. Haseltine and G. C. Pimentel, 'Sinton Bands: Evidence for Deuterated Water on Mars', *Science*, vol. 147 (1 Jan. 1965), pp. 48–9.

Shklovsky, I. S., *Vselennaya Zhizn Razum* (Universe, Life, Intelligence), Press of the Acad. of Sciences of the USSR (Moscow, 1962). See especially Chapter 17, 'Sputniki Marsa – iskusstbennuiye?' ('The Satellites of Mars – Artificial?') English edition, *Intelligent Life in the Universe*, Carl Sagan, ed., Holden-Day (San Francisco, 1966).

Sinton, W. M., 'Spectroscopic Evidence for Vegetation on Mars', *Astrophysical Jour.*, vol. 126 (Sept. 1957), pp. 231–9.

'Further Evidence of Vegetation on Mars', *Science*, vol. 130 (6 Nov. 1959), pp. 1234–7.

'Evidence of the Existence of Life on Mars', *Advances in the Astronautical Sciences*, op. cit., pp. 543–51.

Space Science Board, National Academy of Sciences–National Research Council, 'Biology and the Exploration of Mars – Summary and Conclusions', report by a working group of the board (26 April 1965; mimeo.).

Strughold, Hubertus, 'The Ecological Profile of Mars: Bioastronautical Prospect', *Advances in the Astronautical Sciences*, op. cit., pp. 30–44.

Struve and Zebergs, *Astronomy of the 20th Century*, op. cit., pp. 141–59.

Tikhov, G. A., *Jour. of the British Astronomical Assoc.*, vol. 65 (1955), pp. 193–204.

Young, R. S., P. H. Deal, J. Bell and J. L. Allen, 'Bacteria under Simulated Martian Conditions', *Life Sciences and Space Research*, op. cit.

13. The Uniquely Rational Way

Burke, B. F., 'Contributions of Overseas Observatories', *Physics Today*, vol. 14 (April 1961), pp. 26–9.

Cocconi, Giuseppe, letter to Lovell, 29 June 1959, published as an appendix to *The Exploration of Outer Space*, by Sir Bernard Lovell (Oxford, 1962). In referring to ten planets in the solar system, Cocconi apparently includes the planet that may have disintegrated to form the asteroid belt.

and Philip Morrison, 'Searching for Interstellar Communications', *Nature*, vol. 184 (19 Sept. 1959), pp. 844–6, reprinted in Cameron, *Interstellar Communication*, op. cit., as Item 15.

Flammarion, *La Planète Mars*, op. cit., p. 259 (see p. 328).

General Accounting Office, *Unnecessary Costs Incurred for the Naval Radio Research Station Project at Sugar Grove, West Virginia*, report (Wash., D.C., April 1964).

Jansky, K. G., 'Directional Studies of Atmospherics at High Frequencies', *Proceedings of the IRE* (Inst. of Radio Engineers), vol. 20 (1932), pp. 1920–32.

'Electrical Disturbances Apparently of Extraterrestrial Origin', ibid., vol. 21 (Oct. 1933), pp. 1387–98.

see the *New York Times*, 5 May 1933, p. 1, and 28 June 1933, p. 1.

Ley, Willy, and W. von Braun, *The Exploration of Mars*, Sidgwick & Jackson (1956), p. 7.

Lovell, Sir Bernard, 'Search for Voices from Other Worlds', *New York Times Magazine*, 24 Dec. 1961, pp. 18 ff.

Marconi, G. M., reports strange interference with transoceanic radio traffic. See the *New York Times*, 1920, 27 Jan, p. 7; 28 Jan., p. 5; 29 Jan., p. 1; 30 Jan., p. 18; 31 Jan., p. 24. Reportedly hears signals from Mars. Ibid. 1921, 2 Sept., p. 1; 3 Sept., p. 4.

Pettengill, G. H., H. W. Briscoe, J. V. Evans, E. Gehrels, G. M. Hyde, L. G. Kraft, R. Price and W. B. Smith, 'A Radar Investigation of Venus', *Astronomical Jour.*, vol. 67 (May 1962), pp. 181–90 (on the Millstone Hill observations).

Price, R., P. E. Green, Jr, T. J. Goblick, Jr, R. H. Kingston, L. G. Kraft, Jr, G. H. Pettengill, R. Silver, W. B. Smith, 'Radar Echoes from Venus', *Science*, vol. 129 (20 March 1959), pp. 751–3.

Tesla, Nikola: see 'Nikola Tesla', A Commemorative Lecture by A. P. M. Fleming, given to The Inst. of Electrical Engineers, 25 Nov. 1943; also *Prodigal Genius, the Life of Nikola Tesla*, by John J. O'Neill, Ives Washburn (New York, 1944); also *Return of the Dove*, by Margaret Storm, A Margaret Storm Publication (Baltimore, 1959); also *Electrical Genius – Nikola Tesla*, by A. J. Beckhard, Julian Messner, Inc. (New York, 1959).

quoted by L. I. Anderson in 'Extraterrestrial Radio Transmissions', letter to *Nature*, vol. 190 (2 April 1961), p. 374.

Todd, David, proposal for Mars telescope in Chile: see the *New York Times*, 1921, 7 Sept., p. 2; 8 Sept., p. 17. Gets government cooperation on radio silence: ibid., 1924, 21 Aug., p. 11; 28 Aug., p. 6. Flammarion, aged 82, reaffirms his belief in a superior civilization on Mars: ibid., 22 Aug., p. 13; 24 Aug., p. 30.

Victor, W. K., and R. Stevens, 'Exploration of Venus by Radar', *Science*, vol. 134 (7 July 1961), pp. 46–8 (on Goldstone observations).

R. Stevens and S. W. Golomb, *Radar Exploration of Venus: Goldstone Observatory Report for March-May 1961*, Technical Report No. 32–132, Jet Propulsion Laboratory, California Inst. of Technology (Pasadena, 1 Aug. 1961).

14. Project Ozma

Drake, F. D., 'How Can We Detect Radio Transmissions from Distant Planetary Systems?' *Sky and Telescope*, vol. 19 (Jan. 1960), pp. 140–43, reprinted in Cameron, *Interstellar Communications*, op. cit., as Item 16.

'Project Ozma', *Physics Today*, vol. 14 (April 1961), pp. 40–46.

'Project Ozma', *McGraw-Hill Yearbook of Science and Technology, 1962*, reprinted in Cameron, op. cit., as Item 17.

Emberson, R. M., 'National Radio Astronomy Observatory', *Science*, vol. 130 (13 Nov. 1959), pp. 1307–18.

Struve, Otto, 'Astronomers in Turmoil', *Physics Today*, vol. 13 (Sept. 1960), pp. 22–3.

Thomas, Shirley, 'Frank D. Drake', *Men of Space, Profiles of Scientists Who Probe for Life in Space*, vol. 6, Chilton Books, (Philadelphia, 1963), pp. 62–89.

15. Other Channels

Boehm, G. A. W., 'Are We Being Hailed from Interstellar Space?' *Fortune* (March 1961), pp. 144 ff.

Bracewell, R. N., 'Communications from Superior Galactic Communities', *Nature*, vol. 186 (28 May 1960), pp. 670–71, reprinted in Cameron, *Interstellar Communications*, op. cit.', as Item 25.

'Life in the Galaxy', lecture at Summer Science School, Univ. of Sydney, 8–19 Jan. 1962, published in *A Journey through Space and the Atom*, ed. by S. T. Butler and H. Messel, Pergamon Press (1963); also reprinted in Cameron, op. cit., as Item 24.

Conway, R. G., K. I. Kellermann and R. J. Long, 'The Radio Frequency Spectra of Discrete Radio Sources', *Monthly Notices of the Royal Astronomical Society*, vol. 125 (1963), p. 261.

Drake, F. D., 'The Radio Search for Intelligent Extraterrestrial Life', *Current Aspects of Exobiology*, Pergamon Press (1965).

Dyson, F. J., 'Search for Artificial Stellar Sources of Infrared Radiation', *Science*, vol. 131 (3 June 1960), p. 1667, reprinted in Cameron, op. cit., as Item 11. See also letters in *Science*, vol. 132 (22 July1960), pp. 250–53.

Edie, L. C., 'Messages from Other Worlds', letter in *Science*, vol. 136 (13 April 1962), p. 184.

Golay, M. J. E., 'Note on the Probable Character of Intelligent Radio Signals from Other Planetary Systems', *Proceedings of the I R E*, vol. 49 (May 1961), p. 959, reprinted in Cameron, op. cit., as part of Item 19.

Kardashev, N. S., 'The Transmission of Information by Extraterrestrial Civilizations', *Astronomical Journal* (*Astronomichesky Zhurnal*), vol. 41 (March-April 1964), pp. 282–7. Translation in *Electronics Express*, vol. 6, no. 10 (New York, International Physical Index, Inc., 1964), pp. 37 ff.

Oliver, B. M., 'Some Potentialities of Optical Masers', *Proceedings of the I R E*, vol. 50 (Feb. 1962), pp. 135–41, reprinted in Cameron, op. cit., as Item 22.

'Radio Search for Distant Races', *International Science and Technology*, no. 10 (Oct. 1962), pp. 55–60.

'Interstellar Communication', published in Cameron, op. cit., as Item 29.

Schmidt, Maarten, 'Large Redshifts of Five Quasi-Stellar Sources', *Astrophysical Journal*, vol. 141 (1 April 1965), pp. 1295–1300. His letter is dated 8 April. The journal did not appear until several weeks later, despite its date.

Schwartz, R. N., and C. H. Townes, 'Interstellar and Interplanetary Communication by Optical Masers', *Nature*, vol. 190 (15 April 1961), pp. 205–8, reprinted in Cameron, op. cit., as Item 23.

Sholomitsky, G. B., 'Variability of the Radio Source CTA-102', *Information Bulletin on Variable Stars*, Commission 27 of the I.A.U., No. 83 (Konkoly Observatory, Budapest, 27 Feb. 1965).

Slish, V. I., 'Angular Size of Radio Stars', *Nature*, vol. 199 (17 Aug. 1963), p. 682.

Webb, J. A., 'Detection of Intelligent Signals from Space', Inst. of Radio Engineers Seventh National Communications Symposium Record: *Communications – Bridge or Barrier* (1961), p. 10, reprinted in Cameron, op. cit., as Item 18.

16. Can They Visit Us?

Berosus: see *Cory's Ancient Fragments of the Phoenician, Carthaginian, Babylonian, Egyptian and Other Authors*, by E. Richmond Hodges (London, 1876), p. 57.

Bussard, R. W., 'Galactic Matter and Interstellar Flight', *Astronautica Acta*, vol. 6 (1960), pp. 179–94.

Charles, R. H., ed., *The Book of the Secrets of Enoch*, translated from the Slavonic by W. R. Morfill, Clarendon Press (Oxford, 1896), Chapters 1, 3.

Drake, F. D., 'The Radio Search for Intelligent Extraterrestrial Life', op. cit.

Dyson, F. J., 'Gravitational Machines', published in Cameron, *Interstellar Communication*, op. cit., as Item 12.

Froman, Darol, 'The Earth as a Man-Controlled Space Ship', *Physics Today*, vol. 15 (July 1962), pp. 19–23.

von Hoerner, Sebastian, 'The General Limits of Space Travel', *Science*, vol. 137 (6 July 1962), pp. 18–23, reprinted in Cameron, op. cit., as Item 14.

Pierce, J. R., 'Relativity and Space Travel', *Proceedings of the IRE*, vol. 47 (June 1959), pp. 1053–61.

Purcell, Edward, 'Radioastronomy and Communication Through Space', *Brookhaven Lecture Series*, no. 1, Brookhaven National Laboratory (16 Nov. 1960), reprinted in Cameron, op. cit., as Item 13.

Sagan, Carl, 'Direct Contact Among Galactic Civilizations by Relativistic Interstellar Spaceflight', *Planetary and Space Science*, vol. 11 (1963), pp. 485–98.

Sparks, Jared, *The Works of Benjamin Franklin*, Whittemore, Niles and Hall (Boston, 1856), vol. 8, p. 418.

Spitzer, Lyman, Jr, 'Interplanetary Travel Between Satellite Orbits',

Jour. of the Amer. Rocket Soc., vol. 22 (March-April 1952), p. 93. (Originally published in *Jour. of the British Interplanetary Soc.*)

Tsiolkovsky, K. E.: see letter by A. N. Tsvetikov in *Science*, vol. 131 (18 March 1960), pp. 872, 874. Excerpts from Tsiolkovsky letters were also provided to the author by Dr Tsvetikov.

17. Is There Intelligent Life on Earth?

Blum, H. F., 'Perspectives in Evolution', *American Scientist*, vol. 43 (Oct. 1955), pp. 595–610.

'On the Origin and Evolution of Living Machines', ibid., vol. 49 (Dec. 1961), pp. 474–501.

'On the Origin and Evolution of Human Culture', ibid., vol. 51 (March 1963), pp. 32–47.

Cade, C. M., 'Communicating with Life in Space', *Discovery* (May 1963) pp. 36–41.

Handelsman, Morris, 'Considerations on Communication with Intelligent Life in Outer Space', Paper 4.4, 1962 Wescon Convention Record (Los Angeles, IRE, 21-4 Aug.).

von Hoerner, Sebastian, 'The Search for Signals from Other Civilizations', *Science*, vol. 134 (8 Dec. 1961), pp. 1839–43, reprinted in Cameron, *Interstellar Communications*, op. cit., as Item 27.

Hoyle, Fred, *A Contradiction in the Argument of Malthus, The St John's College Cambridge Lecture 1962–63 delivered at the Univ. of Hull 17 May 1963*, Univ. of Hull Publications (1963).

Lilly, J. C., *Man and Dolphin*, Gollancz (1962).

Morrison, Philip, 'Interstellar Communication', a paper read before the Philosophical Soc. of Wash., 7 Oct. 1960, *Bulletin of the Philosophical Soc. of Wash.*, vol. 16, no. 1 (1962), pp. 59–81, reprinted in Cameron, op. cit., as Item 26.

Morrison, Philip, colloquium: 'Interstellar Communication', Inst. for Space Studies, NASA, 14 Dec. 1961. Stenographic transcript (Ace Reporting Co., Wash., D.C.).

participating in *Proceedings, Radio Astronomy Seminar, N.Y.C.*, 23 Jan. 1962, Amer. Inst. of Physics, pp. 58–101 (mimeo.).

Simpson, G. G. 'The Nonprevalence of Humanoids', op. cit.

18. Celestial Syntax

Bracewell, R. N., 'Life in the Galaxy', op. cit.

Freudenthal, Hans, *Lincos, Design of a Language for Cosmic Inter-*

course, North-Holland Publ. Co. (Amsterdam, 1960), p. 14 and item 4–18–8.

Hogben, Lancelot, 'Astraglossa, or First Steps in Celestial Syntax', an address to the British Interplanetary Soc., 6 Nov. 1952, reprinted in Hogben, *Science in Authority*, Allen & Unwin (1963).

review of *Lincos*, in *Nature*, vol. 192 (2 Dec. 1961), pp. 826–7.

Morrison, Philip, 'Interstellar Communication', op. cit., pp. 78 ff.

see illustrations to 'Are We Being Hailed from Interstellar Space?' by G. A. W. Boehm, op. cit.

Smith, R. F. W., see 'Communication with Extraterrestrial Beings Called Improbable Unless Man Can Signal in Two Systems of Thought', *Science Fortnightly* (30 Oct. 1963), p. 8.

19. What If We Succeed?

Bracewell, 'Life in the Galaxy', op. cit.

Breig, J. A., and Fr. L. C. McHugh, 'Other Worlds – for Man', *America*, vol. 104 (26 Nov. 1960), pp. 294–6.

Cameron, *Interstellar Communication*, op. cit., p. 1 (see p. 316).

Cantril, H., *The Invasion from Mars, A Study in the Psychology of Panic, with the complete script of the famous Orson Welles Broadcast*, Oxford Univ. Press (1940).

Chardin, Pierre Teilhard de, *The Phenomenon of Man*, Collins (1959), p. 286.

Dyson, F. J., letter in *Scientific American*, vol. 210 (April 1964), pp. 8–10.

Harford, James, 'Rational Beings in Other Worlds', *Jubilee* (May 1962), pp. 17–21.

Inge, W. R., *God and the Astronomers*, Longmans, Green & Co. (1933), p. 268.

Jung, C. G., quoted by Harford, op. cit., p. 21.

Macquarrie, J., *The Scope of Demythologizing – Bultmann and his Critics*, SCM Press (1960), pp. 60–61, 179.

Mascall, E. L., *Christian Theology and Natural Science – Some Questions on their Relations*, Longmans, Green & Co. (1956), pp. 36–45.

Michael, D. N., 'Proposed Studies on the Implications of Peaceful Space Activities for Human Affairs', Brookings Institution report to NASA (Wash., D.C., Dec. 1960), pp. 182–4.

Midrash Rabba, commentary on Genesis, Chapter 3, verse 7, Dr H. Freedman, ed. and transl., Soncino Press (London, 1939), vol. 1, pp. 23–4.

Oppenheimer, J. R., 'On Science and Culture', *Encounter* (Oct. 1962), pp. 6–7.

Purcell, Edward, quoted by G. A. W. Boehm, op. cit., p. 146.

Raible, Fr. Daniel C., 'Rational Life in Outer Space?' *America*, vol. 103 (13 Aug. 1960), pp. 532–5.

Russell, Bertrand, *Mysticism and Logic*, quoted in *The Exploration of the Universe*, Part One of *The Citizen and the New Age of Science*, L. B. Young, ed., American Foundation for Continuing Education (Chicago, 1961), vol. 2, p. 368.

The Saddharma-Pundarika or The Lotus of the True Law, transl. by H. Kern, *Sacred Books of the East*, Clarendon Press (Oxford, 1884), vol. 21, Chapters 14 and 15.

Secchi, Fr. Angelo, et. al., see W. D. Muller, *Man Among the Stars*, Harrap (1958), chapter 13 ('Religion in Space').

Shapley, H., *Of Stars and Men*, op. cit.

Smethurst, A. F., *Modern Science and Christian Beliefs*, Nisbet (Welwyn Garden City, 1955), pp. 96–7.

Smith, Joseph, *The Pearl of Great Price*, 'The Book of Moses', Chapter 1, The Church of Jesus Christ of Latter-Day Saints (Salt Lake City, 1952).

Tillich, Paul, *Systematic Theology*, vol. 2, 1953, pp. 95–6, 100–101.
see 'Paul Tillich – Mystic, Rationalist, Universalist', by A. P. Stiernotte, *The Crane Review*, vol. 4 (Spring 1962), pp. 175–6.
quoted in *Newsweek*, 8 Oct. 1962, p. 115.

Wald, George, 'Theories of the Origin of Life', *The Voice of America Forum Lectures, Biology Series*, No. 20, 1960–61. U.S. Information Agency (Washington), p. 7.

Zubek, T. J., 'Theological Questions on Space Creatures', *The American Ecclesiastical Review*, vol. 145 (Dec. 1961), pp. 393–9.

Index